CHEMICAL EVOLUTION
OF THE EARLY PRECAMBRIAN

CONTRIBUTORS

STANLEY M. AWRAMIK
ELSO S. BARGHOORN
G. M. BLISS
W. CALDWELL
W. L. DIVER
RUDOLF EICHMANN
J. FARHAT
CLAIR FOLSOME
SUSAN FRANCIS
P. R. GRANT
E. J. GRIFFITH
D. O. HALL
JOHN B. HENDERSON
THOMAS C. HOERING
R. HURST
HAROLD P. KLEIN
J. LUMSDEN
LYNN MARGULIS
CARLETON B. MOORE
M. D. MUIR
D. W. NOONER
J. ORO
MITCHELL RAMBLER
JOHN J. W. ROGERS
MANFRED SCHIDLOWSKI
J. WILLIAM SCHOPF
BARBARA Z. SIEGEL
RAYMOND SIEVER
E. TEL-OR
JAMES C. G. WALKER
DONNA WELCH
G. WETHERILL
NANCY SYMMES WHITAKER

Chemical Evolution
of the Early Precambrian

Edited by

CYRIL PONNAMPERUMA

Laboratory of Chemical Evolution
Department of Chemistry
University of Maryland
College Park, Maryland

ACADEMIC PRESS   New York   San Francisco   London   1977
A Subsidiary of Harcourt Brace Jovanovich, Publishers

ACADEMIC PRESS, INC.
111 Fifth Avenue, New York, New York 10003

United Kingdom Edition published by
ACADEMIC PRESS, INC. (LONDON) LTD.
24/28 Oval Road, London NW1

Library of Congress Cataloging in Publication Data

College Park Colloquium on Chemical Evolution, 2d,
    University of Maryland, 1975.
    Chemical evolution of the early Precambrian.

    Organized by the Laboratory of Chemical Evolution,
University of Maryland.
    1.   Geology, Stratigraphic–Pre-Cambrian.   2.   Pale-
ontology–Pre-Cambrian.   3.   Geochemistry.   I.   Ponnam-
peruma, Cyril,   Date   II.   Maryland.   University.
Laboratory of Chemical Evolution.   III.   Title.
QE653.C64   1975   551.7'12   77-7165
ISBN 0–12–561360–1

CONTENTS

LIST OF CONTRIBUTORS

STANLEY M. AWRAMIK, Department of Geological Sciences, University of California, Santa Barbara, California 93106

ELSO S. BARGHOORN, Department of Botany, Harvard University, Cambridge, Massachusetts 02138

G. M. BLISS, Department of Geology, Imperial College of Science and Technology, Royal School of Mines, Prince Consort Road, London SW7 2BP, England

W. CALDWELL, Department of Biology, Boston University, Boston, Massachusetts 02215

W. L. DIVER, Department of Geology, Imperial College of Science and Technology, Royal School of Mines, Prince Consort Road, London SW7 2BP, England

RUDOLF EICHMANN, Max-Planck-Institut für Chemie (Otto-Hahn-Institut), Saarstrasse 23, D-6500 Mainz, West Germany

J. FARHAT, Department of Geology and Mineralogy, University of Jordan, Amman, Jordan

CLAIR FOLSOME, Department of Microbiology, Laboratory of Exobiology, University of Hawaii, Honolulu, Hawaii 96822

S. FRANCIS, Department of Biology, Boston University, Boston, Massachusetts 02215

P. R. GRANT, Department of Geology, Imperial College of Science and Technology, Royal School of Mines, Prince Consort Road, London SW7 2BP, England

E. J. GRIFFITH, Inorganic Research Division, Monsanto Company, St. Louis, Missouri 63166

D. O. HALL, Kings College, 68 Half Moon Lane, London, SE24 9JF, England

JOHN B. HENDERSON, Geological Survey Canada, 601 Booth Street, Ottawa, Canada K1A OE8

THOMAS C. HOERING, Geophysical Laboratory, Carnegie Institution of Washington, 2801 Upton Street, N.W., Washington, D.C. 20008

R. HURST, Department of Geology, University of California Santa Barbara, California 93106

HAROLD P. KLEIN, Director, Life Sciences Division, NASA-Ames Research Center, Moffett Field, California 94035

J. LUMSDEN, Kings College 68 Half Moon Lane, London, SE24 9JF, England

LYNN MARGULIS, Department of Biology, Boston University, Boston, Massachusetts 02215

CARLETON B. MOORE, Center for Meteorite Studies, Arizona State University, Tempe, Arizona 85281

M. D. MUIR, Department of Geology, Imperial College of Science and Technology, Royal School of Mines, Prince Consort Road, London SW7 2BP, England

D. W. NOONER, Department of Biophysical Sciences and Chemistry, University of Houston, Houston, Texas 77004

J. ORO, Department of Biophysical Sciences and Chemistry, University of Houston, Houston, Texas 77004

MITCHELL RAMBLER, Department of Biology, Boston University, Boston, Massachusetts 02215

JOHN J. W. ROGERS, Department of Geology, University of North Carolina, Chapel Hill, North Carolina 27514

MANFRED SCHIDLOWSKI, Max-Planck-Institut für Chemie (Otto-Hahn-Institut), Saarstrasse 23, D-6500 Mainz, West Germany

J. WILLIAM SCHOPF, Department of Geology, University of California at Los Angeles, Los Angeles, California 90024

BARBARA Z. SIEGEL, University of Hawaii at Manoa, Pacific Biomedical Research Center, Honolulu, Hawaii 96822

RAYMOND SIEVER, Department of Geological Sciences, Harvard University, Cambridge, Massachusetts 02138

E. TEL-OR, Kings College, 68 Half Moon Lane, London, SE24 9JF, England

JAMES C. G. WALKER, National Astronomy and Ionosphere Center, Arecibo Observatory, P.O. Box 995, Arecibo, Puerto Rico 00612

DONNA WELCH, Center for Meteorite Studies, Arizona State University, Tempe, Arizona 85281

G. WETHERILL, Department of Terrestrial Magnetism, Carnegie Institution of Washington, 5241 Broad Branch Road, N.W., Washington, D.C. 20015

NANCY SYMMES WHITAKER, Planetary Biology Division, NASA-Ames Research Center, Moffett Field, California 94035

PREFACE

We are pleased to offer within the covers of this volume the invited papers (edited and updated) presented at the Second College Park Colloquium on Chemical Evolution held at the University of Maryland from October 29 to November 1, 1975. These meetings have been organized by the Laboratory of Chemical Evolution to promote the interdisciplinary approach to a broadening field of endeavor.

Of paramount importance in the understanding of the processes that led to the beginning of life is the information that can be gathered from a study of the Early Precambrian. In the sediments that were laid down during the very earliest stages of the earth's development may be found a recorded history of pristine evolutionary geology, chemistry, and biology. The latest information available on this subject was critically examined at the Colloquium on the Early Precambrian.

Not only will these studies help us in our attempts to solve the problem "How did life begin?" but we may also be able to pose the question "When did life begin?" in a scientifically defensible manner.

1

ORIGIN OF THE ATMOSPHERE: HISTORY OF THE RELEASE OF VOLATILES FROM THE SOLID EARTH

James C. G. Walker

National Astronomy and Ionosphere Center, Puerto Rico

Terrestrial surface volatiles, which include the atmosphere, the ocean, and volatiles combined in sedimentary rocks, were originally incorporated as compounds in the solid phase of the planet. They were released in a process that we call degassing. Degassing led to the simultaneous growth of the atmosphere, the ocean, and the mass of sedimentary rocks. It presumably accompanied the differentiation of the surface layers of the Earth into crust and upper mantle. This chapter considers whether degassing occurred very early in Earth history or whether it has been a continuous process leading to a gradual growth of the atmosphere. Several lines of evidence are described that suggest that degassing occurred early, possibly during the final stages of accretion of the Earth. It is clear, on the other hand, that the release of gases from the upper mantle to the surface has not been restricted to very early Earth history; indeed, it is probably still occurring. The evidence is not necessarily conflicting. Volatiles may be cycled continuously between the upper mantle and the surface, the return flow being provided by subduction of oceanic sediments. According to this hypothesis, the atmosphere grew, at an early date, to about its present mass. Since that time, recycling has maintained an approximate equilibrium between the mass of surface volatiles and the concentration of volatiles in the upper mantle.

I. INTRODUCTION

Chemical evolution in the early Precambrian must have been

1

significantly influenced by the state of the atmosphere, particularly its composition and surface pressure. I have considered
the composition of the early atmosphere elsewhere (Walker, 1976),
concluding that the principal constituents would have been nitrogen, water, and carbon dioxide in approximately their modern
proportions. As much as 1% of the atmosphere might have been
hydrogen, while oxygen would have been totally negligible.

In this chapter I shall consider the history of the growth
in mass of the atmosphere (proportional to surface pressure) in
order to arrive at a tentative estimate of the surface pressure
in the early Precambrian. This chapter is based largely on material contained in Walker (1977).

II. SECONDARY ORIGIN OF THE ATMOSPHERE

Compared with the sun, the inert gases are depleted on Earth
relative to nonvolatile constituents by amounts so large as to
make it plain that the present atmosphere is not a remnant of
the primitive solar nebula (Moulton, 1905; Aston, 1924; Russell
and Menzell, 1933; Brown, 1952; Sagan, 1967; Fisher, 1976). At
one time or another the materials of the earth were separated
from the nebular gases, retaining only constituents in the solid
phase. It is possible that the Earth was formed by the accretion
of planetesimals after the dissipation of the solar nebula, but
accretion in the presence of the nebula is more likely (Turekian
and Clark, 1968; Clark et al., 1972; Cameron, 1973; Lewis, 1972,
1974). If the Earth achieved approximately its present mass
while immersed in the primitive nebula, its gravitational field
must have attracted a substantial atmosphere of nebular gases.
The rarity of the inert gases on Earth implies that such a primordial atmosphere has been almost completely dissipated. The
only plausible dissipation mechanism that has been suggested is
the T-Tauri solar wind which accompanied the approach of the
sun to the main sequence and which is presumed to have led to
the dispersal of the primitive solar nebula (Hayashi, 1961;
Ezer and Cameron, 1963, 1971).

Thus, any primordial atmosphere of nebular gas that the
Earth might originally have had was lost during the last stages
of formation of the solar system. Chemical evolution in the
early Precambrian therefore occurred in the presence of a secondary atmosphere, formed by the release of gases from the solid
phase of the planet. Degassing is the name given to this release.

This chapter considers the history of terrestrial degassing. There are essentially two points of view on this history (Ozima, 1973). One is that degassing was rapid and early
and has essentially come to a stop (Fanale, 1971). The other
is that degassing has been gradual and continuous and is still

going on (Rubey, 1951).

Direct evidence on the mass of the early Precambrian atmosphere is lacking, of course, but it is not needed. Evidence on the growth of the ocean is equally useful because oceanic and atmospheric growth must both have accompanied degassing. Less obvious, perhaps, is the fact that the accumulation of sedimentary rocks on the surface of the Earth is also a consequence of degassing. Volatiles are absorbed when igneous rocks are converted to sedimentary rocks. In fact, most of the volatiles that have been released by terrestrial degassing reside not in the atmosphere, but in the ocean or in sedimentary rocks. It seems likely, moreover, that degassing accompanied the differentiation of the crust and upper mantle from the unstructured rubble produced by the later stages of Earth accretion.

So we can learn about the history of degassing by seeking evidence on the growth of the ocean, the sedimentary rock mass, and the crust. While no one piece of evidence may be sufficiently compelling by itself to distinguish between rapid initial degassing and continuous degassing, evidence from several sources may possibly yield a convincing degassing history.

In the next section I shall present several arguments that together constitute a strong case for early degassing. There is, however, compelling evidence that volatiles are still being released from the upper mantle to the surface of the Earth. I shall take up this evidence next. Early degassing and continuing release of volatiles are not necessarily contradictory. The apparent contradiction can be resolved by distinguishing between degassing, which results in an addition to the total abundance of surface volatiles (atmosphere, ocean, and sedimentary rocks combined), and release of volatiles from the upper mantle to the surface, which may be accompanied by the return of volatiles from the surface to the upper mantle. This distinction is discussed in the last section of the chapter.

III.   EVIDENCE FOR EARLY DEGASSING

A. Ocean Volume

I present here an argument, due to Armstrong (1968), that the volume of the ocean has not changed in the last 1.7 billion years.

Undeformed, shallow-water sediments ranging in age up to at least 1.7 billion years are found on the stable cratons in the interiors of many of the continents. These sediments indicate that the surface of the sea has been approximately level with the surface of the cratons throughout this period. There is no indication that the crustal structure of the cratons has varied with time; the thickness of crust underlying the cratons shows little variation with position or age. Under the reasonable assumption that oceanic crustal structure has not changed much either, isostasy ensures that the elevation of the craton surface relative to the floor of the ocean basins has been fairly constant. Thus, the average depth of the oceans has not changed in at least 1.7 billion years.

Constant continental thickness and constant ocean depth are possible on an Earth of constant surface area if neither ocean volume nor continental volume have changed with time. Alternatively, one of these volumes may have grown at the expense of the other. The second alternative is not attractive, however, because differentiation of mantle material should cause both continent and ocean to grow together. We conclude that the volume of the ocean has not changed significantly in approximately the last 2 billion years. The same result has been derived from oxygen isotope data by Chase and Perry (1972, 1973). If the ocean has not grown in the last 2 billion years, degassing during this period must have been insignificant.

B. Degassing on Venus

Valuable support for this finding would be provided by a demonstration that the present day degassing rate is negligible. No one has yet been able to come up with an estimate of the present rate of degassing on Earth, but it is possible to set a low upper limit on the rate of degassing on Venus. Since Venus and Earth are quite similar in mass, mean density, and position in the solar system, it is likely that their overall compositions are the same and that they have had similar degassing histories. Their very different atmospheres can be explained entirely in terms of surficial processes (Walker, 1975), without any difference in their degassing histories.

The estimate of the degassing rate on Venus is due to Walker *et al.* (1970). Here I shall present a brief summary of the argument.

The atmosphere of Venus contains no more than 5×10^{20} g of water; any more water would violate the upper limits on the water vapor mixing ratio imposed by measurements of the microwave emission spectrum of the planet (Janssen *et al.*, 1973). At least 2×10^{23} g of water has been released by degassing to the atmosphere of Venus over the lifetime of the planet. This is the amount of water that would have accompanied degassing of the 5×10^{23} g of carbon dioxide in the atmosphere of Venus if carbon was originally incorporated in the planet principally as a hydrocarbon, CH_2. Evidently Venus has lost most of its water, presumably as a result of photolysis followed by escape of hydrogen to space. The reasons for this loss are well understood (Walker, 1975).

Since the abundance of water in the atmosphere of Venus has decreased markedly over the age of the planet it is not likely that this abundance is now increasing. We may therefore conclude that the present rate of degassing of water from Venus is not greater than the present rate of loss of water as a result of photolysis followed by escape of hydrogen. A probable upper limit on the rate of escape of hydrogen from Venus is 10^6 atoms cm $^{-2}$ sec^{-1}, which corresponds to destruction of 2.4×10^9 g of water per year. Degassing at this rate for 4.5×10^9 years would have produced only 10^{19} g of water, a negligible amount compared with the 2×10^{23} g that is a lower limit on the total amount of water degassed from Venus or compared with the 1.35×10^{24} g of water in the terrestrial ocean. The present rate of degassing on Venus therefore appears to be negligibly small compared with rates that must have existed in the past both on Venus and on Earth.

If degassing is negligible on Venus today, it is probably also negligible on Earth. Both the Venus degassing evidence and the ocean volume evidence therefore support a model of fast, early degassing.

C. Lead Isotopes

Early degassing is supported also by data on the lead isotope ratios in crustal rocks. These data indicate that the surface layers of the Earth were perturbed enough to become essentially homogeneous very early in Earth history. Degassing may well have accompanied this perturbation. The implications of the lead isotope date have been described by Turekian and Clark (1975).

Uranium decays radioactively to produce the lead isotopes ^{206}Pb and ^{207}Pb, but not the isotope ^{204}Pb. With the passage of time, therefore, the relative abundances of the different lead isotopes in a rock change in a predictable way at a rate that depends on the ratio of uranium to lead in the rock. Extensive measurements of lead isotopes show that nearly all rocks have had nearly the same uranium-to-lead ratio for nearly all of geological time. Lead isotope data would clearly distinguish any rock that had, at some time in the past, had a uranium-to-lead ratio markedly different from its present value for times as short as 100 million years (Armstrong, 1968). Few such rocks are found.

Uranium and lead are geochemically similar elements, which means that normal geochemical processes are not likely to concentrate one at the expense of the other. Thus, it is not surprising that the uranium-to-lead ratio of most crustal rocks has not changed with the passage of time. What is surprising is that most crustal rocks acquired the same uranium-to-lead ratio very early in Earth history. Uranium and lead are cosmochemically very different, which means that they should have been concentrated in different portions of the materials from which the Earth accreted. Uranium forms refractory compounds that would have condensed from the cooling solar nebula at high temperatures. Lead, on the other hand, forms volatile compounds that would have condensed much later than the uranium compounds. The primitive Earth therefore accreted some material high in uranium and low in lead and some material low in uranium and high in lead. The lead isotope date indicate that the uranium and lead have not been brought together gradually over the lifetime of the Earth. Instead, they were rapidly mixed, probably in the first 100 million years after the Earth accreted. The mixing may well have accompanied and been caused by the last stages of accretion.

Thorough mixing of the surface layers of the Earth may well have been accompanied by degassing and by differentiation of the crust from the upper mantle. The lead isotope data therefore suggest a very early date for the origin of the atmosphere, ocean, and crust.

D. Geological Evidence

Geological evidence in no way contradicts the model of rapid early degassing. Sedimentary rocks were being deposited as long ago as 3.7 billion years (Moorbath et al., 1973), so there must have been substantial ocean, atmosphere, and crust by that time.

Substantial areas of crust older than 2.5 billion years have now been found in the interiors of most of the continents (cf. Windley, 1976). Some petrological evidence suggests that this crust was comparable in thickness to modern continental crust. There is a suggestion that the pattern of tectonic activity in the early Precambrian was different from that in the Paleozoic (Glikson, 1970; Anhaeusser, 1972; Sutton and Watson, 1974), but tectonic patterns depend on the thickness of the lithosphere, not upon the thickness of the crust. The early Precambrian lithosphere may well have been thin as a result of high internal temperatures caused by accretional heating and an initially high level of radioactive heating.

The geological evidence is therefore consistent with an early origin for the crust, atmosphere, and ocean. When the geological evidence is combined with the evidence for constant ocean volume in the last 2 billion years, the evidence for negligible degassing on Venus today, and the evidence for thorough mixing of the surface layers of the Earth at a very early date, a strong case for early degassing emerges.

IV. EVIDENCE FOR CONTINUING RELEASE OF VOLATILES FROM THE UPPER MANTLE

Although the case for early degassing is strong, there is
good evidence that the upper mantle is still rich in volatiles
and that volatiles are still being released from the upper mantle to the surface. I shall summarize this evidence in this
section and propose a possible resolution of the apparent contradiction in the next section.

A. Volatiles in the Upper Mantle

Green (1972) has presented a number of arguments for high
carbon dioxide concentrations in the upper mantle. Petrological
evidence has been presented by MacGregor and Basu (1974). The
structure and mineralogy of kimberlite pipes imply a high content
of volatiles in the source materials, extending to depths of at
least 300 km (Dawson, 1971). The stability relations of carbon
indicate that diamonds form at depths of at least 200 km in an
environment where the partial pressure of carbon dioxide is
approximately equal to the confining pressure of the overlying
rock (Kennedy and Nordlie, 1968). Diamonds contain inclusions of
water, carbon dioxide, and nitrogen (Mitchell and Crockett, 1971).
Roedder (L965) has described inclusions of liquid carbon dioxide
at high pressure in nodules and phenocrysts contained in basalts
derived from the upper mantle.

There is therefore no reason to doubt that volatiles are
abundant in the upper mantle and no reason to suppose that these
volatiles are not released to the atmosphere whenever volcanic
activity brings mantle-derived materials to the surface (Sylvester-Bradley, 1972). In particular, since basaltic magmas origi-
nate in the upper mantle it is likely that basaltic volcanism releases upper mantle volatiles to the atmosphere. There is, however, no estimate of the rate of release of volatiles from the
upper mantle.

B. Inert Gas Evidence

Inert gases in distinctive proportions have been detected
in the rapidly chilled, glassy margins of some midoceanridge
basalts (Dymond and Hogan, 1973; Fisher, 1974, 1976). If, as
seems likely, these inert gases originated in the upper mantle,
the observation constitutes further evidence of the release of
volatiles from the upper mantle to the surface.

It is, moreover, clear from the abundance of radiogenic ^{40}A
in the atmosphere that release of volatiles from the upper mantle
has not been restricted to very early Earth history (Turekian,
1964). Most of the ^{40}A in the atmosphere has been produced by
the radioactive decay of ^{40}K, mainly in the upper mantle. Degassing very early in Earth history could not have provided the
^{40}A in the atmosphere because time was required for the production of ^{40}A by the radioactive decay of ^{40}K. Even with a

high estimate of the K content of the mantle it would have taken 140 million years to produce the A now in the atmosphere. Release of A from the upper mantle was therefore not restricted to the first 100 million years of Earth history (Fanale, 1971). It has probably continued at a significant rate throughout geological time (Turekian, 1964).

V.   DEGASSING AND RECYCLING

I have argued above that degassing occurred early in Earth history. Degassing refers to the accumulation of volatiles at the surface of the Earth which accompanies the growth of the atmosphere, ocean, mass of sedimentary rocks, and probably also the crust. I have also suggested that volatiles are still being released at a significant rate from the upper mantle to the surface. Since this ongoing release is not adding to the total mass of surface volatiles it must be balanced by return of volatiles, at an equal rate, from the surface to the upper mantle (Meadows, 1973; Arrhenius et al., 1974).The only likely locations for the return of volatiles to the upper mantle are subduction zones. Subduction may carry volatile-rich sediments down into the upper mantle.

We can imagine an approximate equilibrium in a cycle of release of volatiles from the upper mantle by volcanism and restoration of volatiles to the upper mantle by subduction. Subduction pumps volatiles into the upper mantle; volcanism releases volatiles when their concentration in the upper mantle rises too high. Thus, it is possible for volatiles to cycle continuously between the surface and the upper mantle without significant change with time in the mass of volatiles in either reservoir.

This hypothetical model renders irrelevant to the question of the growth of the atmosphere a long-standing dispute over the proportion of the volatiles released by volcanoes that are juvenile, meaning that they have never before been in the atmosphere. Volcanic release of volatiles from the upper mantle, whether or not they are juvenile, does not increase the mass of the atmosphere; transfer of volatiles from the upper mantle to the surface by volcanism is balanced by transfer from the surface to the upper mantle by subduction. The rate of release of water from the upper mantle of Venus is small today because water released in the past has been destroyed by photolysis and escape of hydrogen rather than being restored to the upper mantle.

Imagine what would happen on an initially airless planet if the rate of release of volatiles from the mantle were constant. Initially there would be no surface volatiles to be returned to the mantle, so volatiles would accumulate at the surface. As volatiles accumulated, and the mass of sedimentary rocks grew larger, the rate of return of volatiles from the surface to the mantle would rise, and the rate of accumulation of volatiles at the surface would decrease. Equilibrium would be achieved when enough volatiles had accumulated at the surface to sustain a rate of return equal to the rate of release. Attainment of equilibrium would bring to an end the period of degassing; recycling of volatiles between the upper mantle and the surface would continue without further change in the mass of the surface volatiles.

VI. CONCLUSION

Atmosphere, ocean, and sedimentary rocks achieved approximately their present masses in the first 1 or 2 billion years of Earth's history. Indeed, the lead isotope evidence suggests that accumulation of surface volatiles may have occurred within 100 million years of the origin of the Earth, possibly during the final stages of accretion. The surface pressure of the early Precambrian atmosphere, therefore, was probably not very different from the surface pressure of the modern atmosphere. Release of volatiles from the upper mantle to the surface has been balanced by return of volatiles from the surface to the upper mantle since the end of an early period of degassing.

Acknowledgments
    While the conclusions of this chapter are my own, I came across some of the evidence described above at the conference on The Early History of the Earth held in Leicester, England, in April 1975, and I learned of other arguments in discussions with K. K. Turekian. The National Astronomy and Ionosphere Center is operated by Cornell University under contract with the National Science Foundation.

REFERENCES

Anhaeusser, C. R. 1972. *Information Circular no. 70,* Economic Geology Research Unit, University of the Witwatersrand, Johannesburg, South Africa.
Armstrong, R. L. 1968. *Rev. Geophys.,* 6:175.
Arrhenius, G., De, B. R., and Alfven, H. 1974. In E. Goldberg (ed.), *The Sea,* Vol. 5, p. 839. Wiley-Interscience, New York.
Aston, F. W. 1924. *Nature (London),* 114:786.
Brown, H. 1952. In G. P. Kuiper (ed.), *The Atmospheres of the Earth and Planets.* p. 258. University of Chicago Press, Chicago

Cameron, A. G. W. 1973. *Icarus, 18*:407.
Chase, C. G., and Perry, E. C. 1972. *Science, 177*:992.
Chase, C. G., and Perry, E. C. 1973. *Science, 182*:602.
Clark, S. P., Turekian, K. K., and Grossman, L 1972. In E. C. Robertson (ed.), *The Nature of the Solid Earth,* p. 3. McGraw-Hill, New York.
Dawson, J. B. 1971. *Earth Sci. Rev., 7*:187.
Dymond, J., and Hogan, L. 1973. *Earth Planet Sci. Lett., 20*:131.
Ezer, D., and Cameron, A. G. W. 1963. *Icarus, 1*:422.
Ezer, D., and Cameron, A. G. W. 1971. *Astrophys. Space Sci., 10*:52.
Fanale, F. P. 1971. *Chem. Geol. 8*:79.
Fisher, D. E. 1974. *Geophys. Res. Lett., 1*:161.
Fisher, D. E. 1974. In B. F. Windley (ed.), *The Early History of the Earth.* Wiley, New York, in press.
Glikson, A. Y. 1970. *Tectonophysics, 9*:397.
Green, H. W. 1972. *Nature (London), 238*:2.
Hayashi, C. 1961. *Publ. Astron. Soc. Japan, 13*:450.
Janssen, M. A., Hills, R. E., Thornton, D. D., and Welch, W. J. 1973. *Science, 179*:994.
Kennedy, G. C., and Nordlie, B. E. 1968. *Econ. Geol. 63*:495.
Lewis, J. S. 1972. *Icarus, 16*:241.
Lewis, J. S. 1974. *Science, 186*:440.
MacGregor, I. D., and Basu, A. R. 1974. *Science, 185*:1007.
Meadows, A. J. 1973. *Planet. Space Sci., 21*:1467.
Mitchell, R. H., and Crocket, J. H. 1971. *Mineral. Deposita, 6: 392.*
Moorbath, S., O'Nions, R. K., and Pankhurst, R. J. 1973. *Nature (London), 245*:138.
Moulton, F. R. 1905. *Astrophys. J. 22*:165.
Ozima, M. 1973. *Nature Phys. Sci., 246*:41.
Roedder, E. 1965. *Amer. Mineral., 50*:1746.
Rubey, W. W. 1951. *Bull. Geol. Soc. Amer.62*:1111.
Russel, H. N., and Menzel, D. H. 1933. *Proc. Natl. Acad. Sci. U.S.A., 19*:997.
Sagan, C. 1967. In S. K. Runcorn (ed.), *International Dictionary of Geophysics,* Vol. 1 p. 97. Pergamon Press, New York.
Sutton, J., and Watson, J. V. 1974. *Nature (London), 247*:433.
Sylvester-Bradley, P. C. 1972. In C. Ponnamperuma (ed.), *Exobiology, Frontiers of Biology,* Vol. 23, p. 62. North-Holland, Amsterdam.
Turekian, K. K. 1964. In P. J. Brancazio and A. G. W. Cameron (eds.), *The Origin and Evolution of Atmospheres and Oceans,* p. 74. John Wiley and Sons, New York.
Turekian, K. K., and Clark, S. P. 1969. *Earth Planet. Sci. Lett.. 6*:346.
Turekian, K. K., and Clark, S. P. 1975. *J. Atmos. Sci., 32*:1257.
Walker, J. C. G. 1975. *J. Atmos. Sci., 32*:1248.

Walker, J. C. G. 1976. In B. F. Windley (ed.), *The Early History of the Earth*. Wiley, New York, in press.

Walker, J. C. G. 1977. *Evolution of the Atmosphere*. Hafner Press, New York.

Walker, J. C. G., Turekian, K. K., and Hunten, D. M. 1970. *J. Geophys. Res., 75*:3558.

Windley, B. F. 1976. *The Early History of the Earth,* Wiley, New York.

2

EARLY PRECAMBRIAN WEATHERING AND SEDIMENTATION: AN IMPRESSIONISTIC VIEW

RAYMOND SIEVER
Harvard University

Deducing the weathering and sedimentation regimes of the early Precambrian is a disease caused by a virus which is only a slight mutation of that which causes otherwise sensible scientists to calculate the composition of the early Earth's atmosphere and oceans. One can do the job by compounding the assumptions required for atmosphere and ocean with conclusions reached by extracting the fundamental dynamical theories of geomorphology and sedimentation. I will start by restating some of those theories. The nature or topography of a terrain is most simply measured by a combination of altitude and relief, the second being strongly dependent on the first. The topography is a function of two variables: (a) the climate (temperature and rainfall), which affects the nature and degree of erosion, and (b) the ratio of the rates of (tectonism + volcanism) to rates of erosion. I shall return to climate later in this chapter.

The direct dependence of topography on (tectonism + volcanism) is obvious. This relationship also gives a way of measuring the amounts of crystalline (igneous and metamorphic) rocks from the interior that are brought to the Earth's surface and so exposed to reaction with the atmosphere and hydrosphere. Rates of such delivery are conveniently measured by the rates of erosion of the continents, currently a little over 2×10^{16} g/year (Garrels and Mackenzie, 1971). Though volcanism exposes fresh rock directly to the surface, tectonism exposes such rock only through the action of erosion, which gradually lays bare the subsurface as it is elevated. The altitude component of topography is directly related to the rate of erosion; thus, the average heights of the continents of today are directly related to their denudation rates. Over time scales of up to 10^6 years we assume, with geological justification, a steady state between (tectonism + volcanism) and erosion. At time scales of 10^7 years, there is geological evidence of instability, so that for some such periods of time in geologic history the continents may have been higher (as we are today) or lower than the mean value.

Chemical weathering is a function of atmospheric composition; in particular, it is sensitive to the acid gases: CO_2, SO_2, H_2S, and HCL, and to O_2. An early atmosphere significantly different from that of today would have had serious consequences for chemical weathering rates and products. Weathering is also a function of climate, but today the biological world plays an important intermediary role. Much of the effects of temperature and rainfall on rock decay stem from vegetation and bacteria on and in the soil. But significant differences in temperature can effect the kinetics of abiogenic weathering reactions too. Sedimentation kinetics in the ocean can likewise be affected by temperature. In today's world, biology plays a profound role, for example, in restricting the deposition of shallow water carbonate rocks to warmer ocean waters. For these reasons, it is appropriate to digress on the temperature of the early Precambrian Earth.

I. TEMPERATURES OF THE EARLY PRECAMBRIAN

First it is necessary to specify the stage of early Earth history that concerns us. I will start with the initial conditions of an almost wholly acreted Earth, no more than a few hundred million years after the birth of the solar system. A major episode of outgassing had already occurred, supplanting the original gases of the nebula which would have been swept away by the solar wind of the Sun's T-Tauri stage. That outgassing would have been, in part, contemporaneous with melting and differentiation of the interior to form core, mantle, and crust which was probably accomplished in the first 100 - 200 million years of Earth history (Hanks and Anderson, 1969). The consequence was an atmosphere of H_2O, CH_4, CO_2, CO, NH_3, H_2, H_2S, and SO_2 overlying an ocean of water with dissolved salts much like those of today. The last conclusion comes from an analysis of the steady state of the present oceans, whose salts are supplied by the chemical denudation of the continents and depleted by chemical sedimentation. A negligible amount of O_2 might have been present in the atmosphere as a result of photolysis of H_2O in the upper atmosphere.

Perhaps most important for temperature--and yet least susceptible to reliable quantitative estimate--is the value of the solar constant. Most evolutionary models call for somewhat lower luminosity of the Sun at a very early stage, but by the time the Earth was a quarter of an eon old, the Sun would by most models be within 10% of what it is today. In addition to this effect, it is reasonable to look at the other factors determining the surface temperature. Cloud cover, absorption, and back scattering by atmospheric gases might all have been significantly different in the early atmosphere. Our present pressure is presumably in dynamic balance, the major components, oxygen, and nitrogen controlled by the biological photosynthesis - respiration couple. At an early stage of outgassing the atmosphere might conceivably have had a transitory total pressure of up to three to five times that of today, much of it as CH_4. But it is doubtful that by the time which I have selected, the atmosphere would have significantly denser than that of today. The CO_2 would have been generally similar to that of today if the Urey equilibrium between silicate and carbonate were even roughly approximated and p_{NH_3} would have been small, controlled largely by ammonium ion exchange with clays (Bada and Miller, 1968). Hydrogen was almost certainly not above 0.1 atm and probably much lower. So we are left with methane as the dominant gas of that time. In the presence of so much H_2, no likely equilibrium drawdown and regulation of CH_4 at a lower pressure by precipitation of solid compounds seems likely, though upper atmospheric reactions may have destroyed much of the CH_4.

Cloud cover, now so important in radiation balance, is a function of temperature. The water content of the atmosphere is temperature dependent on this water-drenched Earth and is not regulated by mineral hydration (weathering) reactions. Much of the immense amount of water at the surface must have outgassed very early and so I assume a cloud cover at least as much as today and probably a bit higher because of the effect of increased dust.

A higher level of atmospheric dust would have been likely because of the undoubtedly higher level of volcanic activity. Even though the volcanics might have been more dominantly of basaltic and andesitic composition and so had a lower ratio of dust to lava flow, the faster pace of volcanism would have had an important effect. In addition to volcanic dust, the Earth would still have been a partly cratered planet with a significant contribution of macro- and micrometeorites and cratering ejecta dust. It is hard to say what windstorm intensities were in a prebiotic Earth but they must have been able to erode much more dust from the land surfaces unbound by any vegetation. A slight countering effect would be an increased tendency for coagulation and consequent settling of the dust in a denser atmosphere.

Finally, we must consider the reflectance of the land surface, at this time dominated by basaltic and andesitic volcanics,

perhaps slightly darker than the average vegetated soils of
today but probably not greatly different in reflectivity. What
could affect the reflectivity greatly is an ice cover.

Put all of these effects together and I conclude that neither absorption nor bare ground reflectivity would have been
greatly different but that increased dust would have increased
cloud cover and thus increased albedo, decreased ground temperature, and increased polar ice. The ice would have importantly
increased albedo, further lowering surface temperatures. Thus,
an Earth with polar ice caps is a distinct possibility, for it
would take a drop of only a degree or two from today's mean
annual temperature to produce worldwide glaciation. Though it
would be too much to suggest a quantitative radiation budget
for that time, we can establish some limits. The upper bound--
unlikely except at much higher temperatures--would be a runaway
greenhouse effect powered by CO_2 and NH_3. The lower bound would
be the temperature of complete ocean freezeover, though there
would still be smoking volcanoes peeking out of the ice. In
fact, icy oceans surrounding volcanoes might be attractive to
consider as an environment for the origin of life.

II. WEATHERING AT 4.3 BILLION YEARS BEFORE THE PRESENT

With this backdrop of a cool Earth under a primitive atmosphere, we can consider the nature of rock weathering. The major silicates of Group I and Group II metals, typified by the
feldspars, would react qualitatively much as they do today,
depending on the pH of rain water. Even if pCO_2 were not yet
quite as high as today, other acid gases could make up some of
the deficit. Even distilled water can weather feldspar--though
at a slow rate. The presence of NH_3 might have some effect if
one considers the formation of an ammonium clay, for example,
$NH_4Al_3Si_3O_{10}(OH)_2$, the analog of muscovite, by the reaction of
NH_4^+ in aqueous solution:

$$3KAlSi_3O_8 + NH_4^+ + 14H_2O \rightleftharpoons NH_4Al_3Si_3O_{10}(OH_2 + 6Si(OH)_4 + 3K^+ + 2(OH)^-.$$

Weathering of mafic minerals containing reduced metals, mainly Fe^{2+}, would have proceeded much as today, even with a negligible pressure of O_2 in the primitive atmosphere. This is because the rates of dissolution of silicate frameworks of pyroxenes, amphiboles, and other mafic silicates are not limited by
the rate of oxidation of Fe^{2+}, would have proceeded much as today, even with a negligible pressure of O_2 in the primitive atmosphere. This is because the rates of dissolution of silicate
frameworks of pyroxenes, amphiboles, and other mafic silicates
are not limited by the rate of oxidation of Fe^{2+} to Fe^{3+}. Recent
work in my laboratory has shown that, other than oxidation of
small amounts of Fe^{2+} in a thin leached rind, the major oxidation

takes place in the solution, so that there is a relatively high
steady state concentration of Fe^{2+} in the solution surrounding
the dissolving mineral. Mafic minerals are among the most rapidly weathered on today's Earth surface and I believe they
were also in the early Precambrian. I also believe that, be-
carus the rate of oxidation of Fe^{2+} is a function of pO_2, at
pO_2 levels much lower than that of today, but high enough to
make hematite stable, early Precambrian rivers supplied large
quantities of dissolved ferrous iron to the oceans.

A much slower overall rate of silicate weathering, however,
is probable because of the absence of vegetation. Soil formation, which itself promotes chemical weathering, is relatively
rapid, geologically speaking, in the presence of vegetation.
Desert soils of today, even with some vegetation, are thin, slow
to accumulate, and easily eroded by wind. Prebiotic regolith
would have been made cohesive only by the binding action of
moist or dry clays and would have promoted chemical decay only
by its water-holding ability.

Yet, the increased rate of tectonism and volcanism, generating so much fresh surface to reaction, would tend to speed up
the overall chemical denudation rate. What is important to
consider, however, is the ratio of chemical to mechanical weathering. Regardless of the absolute value of chemical weathering,
the overall ratio of chemical to mechanical would have been low,
for in a tectonically active prebiotic Earth, particularly one
at low temperatures, physical fragmentation processes would
dominate. The consequence is a land surface characterized by
youthful and rugged topography, with an abundance of bare rock
and stripped soils and efficient slope wash and gullying leading
to badlands appearance. Ease of eolian erosion and transport
would have led to extensive dune and loess deposits. Extensive
lava and volcaniclastic plains and cones would have covered
appreciable parts of land surfaces. Streams would have tended
to be braided from the oversupply of sediment compared to transport competence, and alluvial fans would have aproned the many
active mountain belts.

Many intermontane basins and craters would have been filled
by playa lakes, perhaps some of them extensive and long lived--
an example from recent geologic history is the Green River Lake
of the Eocene, which lasted at least 2×10^7 years. The nature of
the water in those lakes would have depended much on the evaporation - influx budget, a ratio mainly dependent on aridity and
temperature. Cool, temperate lakes would likely have been fresh.
Arid, warm region lakes would likely have been saline and alkaline.

III.   SEDIMENTATION AT 4.3 BILLION YEARS BEFORE THE PRESENT

The raw material of the sediments of the Early Precambrian would have been the abundant immature unweathered detritus of preexisting rocks--mainly crystallines--and fresh volcanics. No doubt the ratios of gravel to sand and sand to clay were higher than today, for it is chemical decay that tends to form the fine sizes, and true clay sizes are mainly of the clay minerals, which come only from chemical weathering. In other words, the major sediment types would have been arkoses and subarkoses on the continents and nearshore, and feldspathic and volcanic graywackes of the turbidite class in the deeper continental borderlands and ocean basins. Volcaniclastics and gradations between them and terrigenous sediments would have been abundant.

The shorelines of the oceans, qualitatively of the same morphologic varieties as today, would have had many more and larger coarse-grained detrital sedimentary accumulations, such as deltas, beaches, barrier islands, and shallow sand banks at or near points of major river entry. At the same time, rugged tectonic and volcanic shorelines with rocky headlands and small bays would also have been important. What may well have been less important were the extensive coastal plains of today, bordered by long strips of shallow shelf - beach - barrier - lagoon salt - marsh complexes typical of that regime. Rather, the continental borders and shelves would be narrow and tectonically defined, perhaps similar to those off California today. The deeper continental borderlands and their sedimentary basins would have been filled with coarse turbidites of volcanic graywacke composition, dissimilar to those of today only in the rates of sedimentation and the proportion of volcanic glass.

Chemical sedimentation patterns would have been far different from today. First, I consider limestones, now the dominant chemical sediment and largely precipitated as shell material followed by fragmentation, physical dispersal, and sedimentation. Without biological interference an ocean could be expected to come to a steady-state equilibrium with the influx of Ca^{2+}, Mg^{2+}, and HCO_3^- supplied by streams from the weathering of the continents. In an ocean at a slightly lower temperature, we would expect slightly more dissolved CO_2 and an increased solubility of carbonate so that the equilibrium Ca^{2+} and HCO_3^- levels would have been slightly higher than those of today. In such an ocean, perhaps supersaturated by a factor of 2 or 3, carbonate precipitates would no doubt have included much aragonite, the polymorph whose kinetics of nucleation and crystal growth are fastest.

It is highly unlikely that such a precipitate would have been thinly spread over all of the ocean basins; the supersaturation level needed for precipitation would have been reached easily and first in slightly hypersaline, warmed, shallow near-shore waters--roughly the same kinds of environments as today but with a significant difference: no wave-resistant biogenic structures such as reefs to create the back reef lagoons and shallow banks that we associate with the carbonate sedimentary environments of today. In a prebiotic world there would be ev-ery reason to expect all degrees of mechanical mixtures of a fine-grained carbonate and terrigenous mud in the offshore areas and perhaps carbonate-coated sand grains in the surf zones. Re-latively pure limestones would be localized at places with little or no influx of terrigenous detritus, likely to be low-lying swampy coastlands of moderate subsidence and absence of volcanism, relatively rare places on the early Earth.

Almost the same conditions could be predicted for evaporite deposits of this time, for in some sense the carbonates are the first evaporites to come out. Whenever localized high evapora-tion rates were so high that, as carbonate precipitation (inclu-ding magnesian calcites and dolomite) continued, the solubility product of gypsum was exceeded sufficiently, $CaSO_4 \cdot 2H_2O$ would have been deposited. There is no reason that the other saline precipitates would not have followed in some proportion of these locales. Yet, we have, so far, little evidence of the collapse breccias that might be relics of an extensively dissolved evapor-ite sequence. I suggest later slow dissolution and replacement because I subscribe to the view that the well-known decrease in relative amounts of limestone and evaporite in the Precambrian is a matter of differential preservation (Garrels et al.. 1972). Carbonate and evaporite may have easily dissolved during the billions of years of subsequent geologic history, typically to be replaced by silica. Once silica is emplaced, it is so slow to dissolve and so slightly soluble that it remains, either pre-cipitated at first or ultimately recrystallized as quartz, one of the most stable materials at the Earth's surface.

The only proviso for the above deduction is the probability of sufficiently high evaporation rates, presumably in the warmer equatorial belts. Were surface temperatures to have been so cold that evaporation rates and regional inhomogeneities were greatly depressed, a different scheme might have to be envisioned, in which the general, more uniformly distributed evaporation of the world ocean surface would have controlled precipitation. In such a case we would have to expect that carbonate and gypsum would have been widespread oceanic sediment components. Sooner or later, Na^+ and Cl^- would have built to the point, probably far short of saturation with respect to halite, where they were re-moved by cation and anion exchange with clays, primarily the montmorillonites.

We now come to the most interesting chemical sediment: silica. Interesting because of its association in the Precam-

brian with fossil algal and bacterial remains and because of
its widespread occurrence and unusual lithologic association with
iron formations. In the absence of silica-sequestering organisms, we can consider the balance between supply of dissolved
silica from both continental weathering and marine hydrolysis of
volcanic glass and precipitation of silica as free silica, or
by reaction with, or sorption on, silicates to form reconstituted
clay minerals or zeolites. In a well-mixed ocean at a pH of 8,
reactive soil clays could buffer the silica by moderately rapid
sorption to levels between 15 (kaolinite) and 35 (montmorillonite)
mg/liter ($10^{-3.2}$ - $10^{-3.6}$ M) in SiO_2 (Siever and Woodford, 1973).
Only at much lower pH values could the oceans be buffered at high
silica concentrations, for example, about 100 ppm ($10^{-2.7}$ M) for
montmorillonite at a pH of 5.5. But measured rates of sorption,
though high, are strong functions of clay surface areas. The
clay provided by the suspended sediment reaching the oceans,
approximately 10^{16} g/year currently, would not have been able
wholly to sorb the influx of silica, at least 10^{14} g/year at that
time, for that would imply a sorption of at least 1% by weight,
highly improbable if surface areas were in the vicinity of 5 - 25
m^2/g for most terrigenous clays.

So I conclude that amorphous silica precipitation was no ac-
cident but the expectable way in which the oceans got rid of a
fair share of their silica influx. I would guess that the oceans'
dissolved silica would have been rate-controlled by a combination of silicate reactions and amorphous silica precipitation
at open ocean levels of about 40 - 60 mg/liter, somewhat lower
than the solubility of amorphous silica at 0°C. I again involve
the evaporite model for localized areas where dissolved silica
could have built up to amorphous silica saturation levels: some
combination of high evaporation rates, poor mixing with the open
ocean, relatively low influxes of clays, and high amounts of
volcanic glass. The chert pseudomorphs of halite found in the
Gunflint Formation are evidence for this association to a later
time (Barghoorn, personal communication). In such environments I would expect silica precipitation from only slightly
supersaturated hypersaline seawater to give a finely particulate
material that could be transported by currents. It would be
"gelatinous" only to the extent that the individual grains
would have included much water, much as clay floccules do in
similar environments today. The result: bedded chert.

Now envision this silica regime after procaryotic algae had
evolved. Any algal mats of buildups could form silica stromatolites in an analagous fashion to the well-known carbonate
stromatolites of the Precambrian and today. I have no doubt
that some were combined silica-carbonate originally, but that
the vicissitudes of leaching and replacemetn by silica have re-
moved much of the carbonate from the more ancient ones. Commonly found carbonate inclusions may bear witness to a former mix-

ed composition (Barghoorn and Tylae, 1965). The balance between
carbonate and silica may have depended largely on volcanic activity, carbonate more common in nonvolcanic places and intervals
and silica in some proximity via water activity to volcanic environments. Obviously, the many pure Precambrian cherts that
show only small amounts of terrigenous of volcanic detritus must
have formed in environments that have been called "starved" of
detrital supply, either by trench-type deep water barriers or
because their hinterland was largely limestone, as is the case
with many carbonate environments of today It is also tempting
to suggest extensive deep-sea floor volcanism just seaward of a
protected shelf with little detritus.

There are grounds for believing that in camparison to today,
there was a larger total amount of dissolved silica that entered
the oceans in the early Precambrian, and therefore a larger proportion of the annual silica budget that was precipitated a free
silica. My reasons for suggesting this is that the absolute amount of chemical weathering was probably higher than that of today. Even though mechanical weathering was dominant, the total
amount of chemical weathering would have been greater because of
the abundance of fresh plutonic and volcanic rocks presented to
the atmosphere for reaction. If we take feldspar weathering as
representative, approximately 2 mol of dissolved SiO_2 are released for every mole of feldspar decomposed by 1 mol of carbonic
acid. If we estimate that the rate of [tectonism + volcanism]
and thus the quantity of rock weathered, was greater by a factor
of 3 or 4, then the increase in dissolved SiO_2 might have ap-
proached an order-of-magnitude greater influx than that of today.
The fraction precipitated as free silica would have increased
correspondingly. If that increase were to be spread, not over
the entire oceans, but concentrated in nearshore and estuarine
zones, then the apparent increase in chert proportion of the
stratigraphic sections preserved would be dramatic.

Not the least important for those of us interested in how
and where life started is the matter of organic sedimentation.
Before life, the partition between organic and carbonate carbon
would have been very different from that of today and dependent
on the oxidation state of the atmosphere, ultimately traceable
to atomic hydrogen escape from the upper atmosphere. Nonequilibrium processes are always common on Earth, and I suggest that
even at a time when hydrogen had dropped to levels where equilibrium carbonate precipitation could take place, precipitation
of solid organic material would still have taken place, primarily
the products of ultraviolet and electrical discharge synthesis
of complex mixtures of organic compounds. The bulk of this material would have been yellow tars or "dark brown matter" that is
produced along with amino acids and other extractable labile compounds. Coagulation, particle growth, and physical sedimentation
of such reduced carbon would have contributed a small but steady

heavy organic carbon fraction to sea-bottom sediments, a conclusion amply justified by studies of sedimentation of modern
residual oil spills in the ocean.

Would such organic carbon deposits have been localized, and
if so, where? We can conservatively assume that the bulk of the
organic synthesis took place by sorption of uv light in the thin
surface layers of the ocean, with additional yields from lightning and corona discharge in the atmosphere. The organic products
of this primary synthesis would have been more or less uniformly
distributed over the oceans by both routes of synthesis; the
amounts from electrical discharges over land would have produced
some fallout over the continents.

Though many of the labile components would have decomposed
quickly, the difference between rates of production and decomposition in a cold ocean would have led to a constant steady-state
concentration averaged over the whole oceans. If we grant that
the oceans, then, as now, circulated horizontally and vertically
in patterns governed by the latitudinal temperature distributions,
wind distributions, and the Earth's rotation, we can expect that
the oceans would have been rotating in large gyres with warmer
and colder boundary layer currents at their borders, as the Gulf
Stream and Humboldt current are today. Since decomposition is
a function of temperature, we would expect that some slow advec-
tion of soluble organic compounds would tend to concentrate amino
acids and other compounds in very cold waters at high latitudes
subject to intermittent freezing. The oceans' movements could
thus have been responsible for sequestering the raw material of
precellular polymerization where it could do the most good. The
dark brown matter would have been buried in bottom sediments,
thus initiating in the prebiotic world the subsurface part of
the carbon cycle. This burial of carbon would have been the
only contribution of the surface to reduction in the interior
that was not balanced by oxygen production in the atmosphere,
as became the case after photosynthesis came about.

The last but not least question of chemical sedimentation
in the early Precambrian is the origin of iron formation. That
subject is much too loaded with controversy and generations of
geologists' and geochemists' proposals for me to cover meaning-
fully in this brief, impressionistic survey. Suffice it to say
that iron formations (not restricted to minable ore depostis)
extend as far back in the Archean rock record as we can go and
probably to the time I am here considering. At the same time,
known Precambrian iron formations are so diverse in oxidation
state, minerology, and lithologic associations that one sedimentary environment could not possible account for all types. After
we come to a better understanding of later Precambrian iron formations, we can start to think of the earlier ones.

REFERENCES

Bada, J. L., and Miller, S. L. 1968. *Science, 159*:423.

Barghoorn, E.S., and Tyler, S. A. 1965. *Science, 147*:563.

Garrels, M. and Mackenzie, E. T. 1971. *Evolution of Sedimentary Rocks,* Norton, New York.

Garrels, R. M., Mackenzie, F. T., and Siever, R. 1972. In Eugene C. Robertson (ed.), p. 93. *The Nature of the Solid Earth,* p. 93. McGraw-Hill, New York.

Siever, R., and Woodford, N. 1973. *Geochim. Cosmochim. Acta, 37;* 1851.

3

VERY OLD (> 3100 MILLION YEARS) ROCKS IN NORTH AMERICA

R. HURST
University of California, Santa Barbara

J. FARHAT
University of Jordan

G. WETHERILL
Camegie Institution of Washington

Rb - Sr and U - Pb age measurements have been carried out in the Nain Province of Labrador and in the Minnesota River Valley.

Geological similarities between Saglek Bay, Labrador and Godthaab, West Greenland suggest that some Precambrian Gneisses in Labrador may be equivalent in age to the very old ($\sim$3700 million years) rocks reported by Moorbath and co-workers in Greenland. Subsequent geological mapping in the Saglek area by Bridgwater, Collerson, and Hurst resulted in a tentative identification of specific rock units with those in Greenland. Whole rock Rb - Sr measurements of the older gneiss suite gives an age of 3622 ± 72 million years, and an initial $^{87}Sr/^{86}Sr$ ratio of 0.7014 ± 0008 (2σ). The younger undifferentiated gneisses yield an age of 3120 ± 160 million years, 0.7064 ± 0.0012. These results support the hypothesis that the Uivak Gneisses may be identified with the Amitsoq Gneisses of Greenland and the undifferentiated gneisses with the younger Nuk Gneisses. In addition, layered Archean anorthosite complexes found in the Okhakh Island-Tessiuyakh Bay area may be the equivalent of the Fiskenaesset anorthsites of Greenland. Less complete measurements in the latter region suggest that very old Archean Gneisses occur there also. Barton has reported 3600-million-year-old gneisses (Rb - Sr whole rock) from the vicinity of Hebron Bay. Taken altogether, these data suggest an extensive region of very old rocks in Labrador, which probably were contiguous with those of West Greenland prior to Tertiary North Atlantic rifting.

Whole rock Rb - Sr and U - Pb measurements on zircons from

the Morton and Notevideo Gneisses in the Minnesota River Valley show that this is another area of North America in which >3100-million-year-old rocks are exposed. However, in this region, pervasive metamorphism 2400 - 2600 million years old largely reset isotopic radiometric systems, causing it to be very difficult to uniquely identify the original age(s) of these rocks. U - Pb date suggest an age of >3300 million years, whereas Rb - Sr data clearly demonstrate only that the original age of crystallization was approximately >3100 million years. Very old Rb - Sr isochrons (∿3800 million years old) reported by Goldich and coworkers of ∿3800 million years could not be reproduced in our work, which suggests that their result could be an artifact caused by incomplete sampling of badly disturbed isotopic systems. It is concluded that while it is clear that these rocks are among the oldest yet found in North America, their similarity in age with those of Greenland and Labrador remains to be demonstrated.

REFERENCES

Farhat, J. S. 1976. in preparation.

Farhat, J. S., and Wetherill, G. W. 1975. *Nature (London)*, *257:* 721.

Hurst, R. W., Bridgwater, D., Collersen, K. D., and Wetherill, G. W. 1975. *Earth Planet. Sci. Lett.* *27*:393.

4

THREE ARGUMENTS FOR CONTINUAL EVOLUTION OF SIAL THROUGHOUT GEOLOGIC TIME

JOHN J. W. ROGERS

University of North Carolina, Chapel Hill

Many theories for the evolution of continental material envision segregation of most presently existing sial during the early stages of earth history, with later orogenic events largely reworking this initial material. The early formation of the continents simplifies the problem of explaining the apparent constancy of sea level relative to continental elevation throughout much of geologic time. There are, however, three observations that indicate that considerable new sial has segregated out of the mantle at virtually all times in earth history. First, some continental shield areas appear to have formed in Proterozoic time and do not show evidence of reworking of earlier Archean continental crust. Examples of these shields may be the Llano uplift of central Texas and the Eastern Desert of Egypt. Second, Phanerozoic orogenic belts now incorporated within continents can be shown to have evolved from initially oceanic areas. An example is the Cordilleran belt of western Nevada and eastern California, where volcanic and associated sedimentary features show westward growth of the North American continent through much of Late Paleozoic and Mesozoic time. Third, graywackes and associated sediments of geosynclines formed during all ages contain appreciable quantities of debris not derived by the erosion of well-formed cratons. The extremely low potassium feldspar/plagioclase ratio in these rocks indicates source materials similar to those now occurring in island arcs. These materials are not incorporated in continental cratons, which presumably attained their present composition after the geosynclinal sedimentation.

I. INTRODUCTION

    One of the major concepts of modern geology is the evolu-
tion of most, perhaps nearly all, of the Earth's continental,
sialic material in early Precambrian time. This early evolution, and later reworking, of sial has been proposed on the
basis of a variety of evidence, including: (a) apparent ensialic development of Proterozoic and Paleozoic rocks based
on their comparatively high K_2O/Na_2O ratios (Engel et al.,
1974); (b) apparent incorporation of crustal-derived Sr and
Pb isotopes into orogenic rocks of all ages (Armstrong and Hein,
1973): and (c) the possibility of granitic basement beneath
Archean, and possibly younger, greenstone belts (e.g., McGlynn
and Henderson, 1972: Hunter, 1974a,b). These observations have
been specifically challenged in a number of papers, such as:
(a) Mitchell (1975) discusses the possibility of ensimatic
activity throughout geologic time; (b) a number of studies has
shown mantle Sr and Pb isotope ratios to be characteristic of
island arc, orogenic rocks (e.g., Donnelly et al., 1971); and
(c) much of the granitic "basements" of many greenstone belts
may actually be younger than the greenstones (e.g., Anhaeusser,
1973; Glikson and Lambert, 1973).
    Perhaps the most compelling argument for maintenance of
a constant volume of continental sial throughout much of
geologic time is the observation that shallow-water sediments
have formed at all ages since the early Archean. This observation is particularly strengthened by the occurrence of numerous stromatolites in the Proterozoic and a few in the Archean
(e.g., Henderson, 1975), although C. Stock (personal communication, 1976) indicates that most of these stromatolites are on the
margins of age provinces and, therefore, may not represent
major marine incursion onto cratons. The continued formation
of shallow-water sediments presumably indicates relative constancy of sea level with respect to average continental elevation
since the early Precambrian.
    Because sea level at any time is related to such major
features as the radius of the earth, the volume and thickness
of continents, the volume of sea water, etc., it is clear
that variation in any one of these properties (such as the volume of continents) requires delicate variations in one or more
of the other variables in order to maintain essentially constant
sea level (Wise, 1974). The simplest, and therefore the most
attractive, method of maintaining a balance between the variables that control sea level is a lack of variation in any of
the variables. A constant sea level is obviously maintained
if there has been no change in the volume and thickness of
continents, volume of sea water, etc., since the early Precambrian, with the repeated incursions and withdrawals of

epicontinental seas merely representing minor fluctuations in some property.

This chapter discusses the possibility of variation in the volume of continental sial throughout geologic time. It proposes that new sial has been added to the crust almost continually since early in Precambrian time and bases this conclusion on three arguments: (a) the development of new shields during Proterozoic time that are similar to shields formed during the Archean; (b) evidence of continental extension into oceanic areas during the Phanerozoic; and (c) the continual development of graywackes and associated sediments from primitive source rocks not characteristic of well-formed shields and the later incorporation of these sediments into continental cratons.

II. PROTEROZOIC SHIELDS

Vast areas of Precambrian shield record the 2500 - 2600-million-year (Kenoran) event as the last major tectonic or thermal episode. Other areas, however, are dominated by Proterozoic ages extending to as young as the approximately 500-million-year Pan African (Pangeaic) orogeny. The Proterozoic events in the Canadian shield (see Stockwell, 1968) are commonly considered to represent reworking of older crust on the basis of relict ages and structures found in younger provinces and some correlation of units across province boundaries (e.g., see Wynne-Edwards, 1972, for the Grenvill Province). Paleomagnetic evidence (Donaldson *et al.*, 1973) shows that most of the present Canadian shield was coherent from the end of the Archean to the present, with the possible exception of independent movement of part of the Grenville Province about 1200 million years ago.

One principal indication that not all Proterozoic events represent reworking of older crust is the work of Hurley (1972 1974) to the effect that Pan African (Pangeaic) belts of 650 - 250 million years of age contain abundant sial newly derived from the mantle. This conclusion is largely based on Sr^{87}/Sr^{86} isotope ratios in the vicinity of 0.705, which preclude derivation from materials that have had any appreciable residence time in rocks of crustal Rb/Sr ratios.

In this regard, two shield areas may be of considerable significance. The first is the small, Precambrian, Llano uplift of central Texas, U.S.A., and the other is the Eastern Desert of Egypt.

The major rock units of the Llano uplift are the basal Valley Spring Gneiss, overlying Packsaddle Schist, and a series of late intrusions (primarily Town Mountain Granites). Limited dating has placed the regional metamorphism and intrusion of the Town Mountain Granites at about 1000 million years, with whole-rock samples of the Valley Spring Gneiss yielding 1120 million-year ages (Zartman, 1964, 1965). No structural or isotopic evidence exists for overprinting of earlier rocks or derivation of the granites by remelting of former sialic crust, although the Llano area has been considered to be an extension of the Grenville belt (Muehlberger *et al.*, 1967), in which such remetamorphism has been identified.

The Llano uplift is similar to Archean shield areas in its sequence of development of major rock types, namely: old sialic gneiss (Valley Spring) overlain by more mafic, greenstone-like metasediments (Packsaddle) and intruded by post-tectonic granites (Town Mountain). The average composition of the Llano uplift (Table 1) is also similar to that of the Canadian Archean, the only major difference being the higher K_2O content of the Llano.

TABLE 1
Comparison of Llano and Canadian Archean

	Llano uplift (%)[a]	Canadian Archean (%)[b]
SiO_2	70.7	65.1
TiO_2	0.35	0.50
Al_2O_3	13.6	16.0
ΣFe_2O_3	3.4	4.8
MgO	1.1	2.3
CaO	2.6	3.4
Na_2O	3.3	4.1
K_2O	4.4	2.7

[a]Llano uplift modified from Johnson (1975).
[b]Canadian Archean from Eade and Fahrig (1971).

The Precambrian shield of eastern Egypt (summary by Said, 1972) shows many similarities to the Llano area. The sequence of major rock types in Egypt is: (*a*) old, sialic, gneiss; (*b*) greenstone-like metasedimentary and metavolcanic suites apparently overlying the gneiss; (*c*) intrusion of the metamorphic rocks by an older sequence of gray granites; (*d*) later intrusion by a series of pink granites (Younger Granites); and (*e*) development of postcratonization volcanic and shallow-water sedimentary assemblages. Dating by Hashad *et al.* (1972) and El Shazly *et al.* (1973) places the age of the metamorphic rocks at 1150 to 1300 million years, of the Older Granites (gray) at 850 to 1000 million years, and of the Younger Granites at 500 to 600 million years. These dates are similar to, but about 100 million years younger than, the dates of equivalent rock types in the

Arabian shield (Fleck, 1972). Evidence of reworking of older
crust is largely missing from the Egyptian Precambrian.

Indications of similarity between the young (Llano and
Egyptian) shields and Archean shields are strengthened by the
comparison of average compositions of shields shown in Table 1
for two areas where data are available. Furthermore, preliminary information indicates that compositions of individual rock
types are comparable between old and young shields. For example,
Packsaddle metasediments are compositionally similar to para-
gneisses and paraschists from the Archean of the Superior Province of Canada, to "metamorphites" in the Ancient Gneiss Complex
of Swaziland, and to metasediments fron the Dharwar greenstone
belts of India. Similar comparisons can be made for the Valley
Spring Gneiss and Town Mountain Granites, but comparable data
for the Egyptian Precambrian are lacking.

The only apparent difference between old and young shields
is the proportions of rock types that characterize them. Table
2 shows generalized percentages of major rock types in selected
shield areas obtained by point counting of geologic maps. Considering the great differences between designations of lithologic
units in different areas, it is impossible to attach much significance to the data. It seems apparent, however, that the young
shields contain a higher proportion of young, potassic granites
than do their Archean equivalents, and it is also notable that
granulite-facies rocks are not reported either from the Llano
uplift or from eastern Egypt. The higher proportion of potassic
granites is the reason for the slightly more potassic composition
of the Llano area as compared with the Canadian shield. It is
unclear whether these differences in rock type are related to
shallower depths of erosion in young shields or to some other
factor.

The major conclusion to be drawn from the preceding discussion is that shields similar to Archean shields can evolve
at comparatively young times in the Earth. The two examples
used here, the Llano uplift and the Eastern Desert of Egypt,
apparently represent cratonization in the general range of
500 to 1000 million years ago. Present knowledge of the geology of North Africa and the basement of North America does
not permit determination of the areal extent of these young
provinces. The conclusion, however, seems inescapable that
some considerable amount of sial has been freshly derived from
the mantle during post-Archean time by processes similar to
those that caused the formation of the Archean shields.

III.   PHANEROZOIC CONTINENTAL GROWTH

Additional evidence for the continual evolution of sial
throughout geologic time is the observation of Phanerozoic con-

TABLE 2
Percentages of Rock Types in Various Shields [a]

	Llano uplift (%)	Eastern Egypt (%)	Canada (Superior Province) (%)	India (%)
Young, potassic Granites	30	13	0.4	--
Old Granites to granodiorites	17	28	24	8
Mafic metasediments and metavolcanics	17	34	8	18
Gneisses and migmatites	35	8	52	66
Granulites	-	-	14	8
Probable age of cratonic stabilization	1000 million years	500 million years	2500 million years	2500 million years

[a] Percentages of rock types in the Superior Province are calculated from Eade and Fahrig (1971) and are based on point counting of geologic maps for the other areas.

tinental growth. Documentation of this growth requires the
location of former continental margins and the mapping of their
migration with time.

One major effort of this type has been made in the Western
United States (summary by Rogers *et al.*, 1974). In this area,
volcanic rocks are common constituents of the Paleozoic and
Lower Mesozoic sequences. These volcanic suites can be subdi-
vided on the basis of chemical properties into four environments of formation: (a) continental, including potassic (sho-
shonitic by some definitions), rocks extruded through a normal
continental crust; (b) continental margin, characterized by
calcalkaline assemblages of the Andean type of Jakes and White
(1972); (c) island arc, including tholeiitic and calcalkaline
assemblages found in ensimatic island arcs: and (d) oceanic
or behind-the-arc basin suites recognized by their preponderance
of tholeiities similar to midocean-ridge basalt.

Volcanic rocks characteristic of these various environments
show both lateral and vertical patterns of distribution in
western Nevada and eastern California. Essentially, at any
one time, oceanic and island arc rocks are formed in the west,
continental rocks in the east, and continental margin assemblages in between. Also, at any one place, oceanic and arc assemblages tend to grade upward into continental margin and continental suites. Thus, the continental margin has moved westward
during late Paleozoic and early Mesozoic time. The entire area
is now underlain by continental crust (somewhat attenuated in
the Great Basin), and the best interpretation of the sequence
of environments and present condition is a progressive enlarge-
ment of the continent westward into areas formerly occupied by
oceanic crust. This movement implies continued segregation of
sial into previous oceanic areas during late Paleozoic and
Mesozoic time.

The source of the new sial in western North America is not
certain but is presumed to be the upper mantle. The principal
reason for this conclusion is the volumetric consideration that
the new sial represents approximately 5% of the western part
of North America. Derivation of this sial by erosion of the
preexisting North American craton, incorporation of the eroded
material into the orogenic belt and upper mantle, and remobili-
zation to form the new continental crust should leave some evidence of extensive crustal thinning elsewhere in western North
America; this evidence is apparently missing. Derivation of
the new sial by similar processes acting on other preexisting
cratons is also unlikely because the late Paleozoic and early
Mesozoic were times of maximum continental aggregation, with
closure of the Atlantic and Indian Oceans, thus indicating the
absence of any other craton near the western margin of North
America.

An additional example of Phanerozoic continental growth may be found in Indonesia. At the present time, the major islands of Indonesia occur on the Sunda Shelf, an extensive area of shallow water apparently underlain by crust intermediate between continental and oceanic and possibly partly sialic (based on geophysical data of Ben-Avraham and Emery, 1973). The islands, however, consist primarily of oceanic and ensimatic island arc assemblages formed during the late Paleozoic, Mesozoic, and Tertiary (summary by Katili, 1975). Thus, it is not unreasonable to assume that the partially sialic (?) material of "Sundaland" represents a recent segregation into a formerly oceanic area, although a great amount of detailed work will have to be done in the area in order to provide further documentation.

As mentioned above, approximately 5% of western North America appears to have formed in late Paleozoic and early Mesozoic time. The period of time involved is roughly 200 million years, which is approximately 5% of the age of the Earth. This calculation, plus the general evidence summarized in this section, make it tempting to conclude that new sial segregates out of the mantle at a roughly constant rate throughout geologic time; but obviously, this conclusion must be only a suggestion until a great deal of additional evidence is produced.

IV.   COMPOSITION OF GRAYWACKES

Graywackes and associated sediments are among the major components of geosynclinal assemblages. Commonly, they are associated with volcanic rocks to form eugeosynclinal assemblages characterized by turbidite deposition, great thickness of accumulation, and considerable stratigraphic and structural complexity. The significance of their chemical and petrographic composition has been discussed by a number of writers, including Condie et al. (1970), Pettijohn (1970, 1972), Henderson (1972), and Rogers and McKay (1972).

Graywackes have formed in geosynclinal belts of all ages. They are major components of Archean greenstone belts, form great thicknesses in certain Proterozoic terranes (e.g., the Southern Province of the Canadian shield, Card et al., 1972), are present as major portions of certain Late Proterozoic shields (e.g., the Llano uplift and eastern Egypt, discussed above), occur in Paleozoic geosynclinal belts, including the type graywacke of the Harz Mountains, Germany (e.g., Helmbold and Van Houten, 1958), and are very abundant in continental margin and island arc orogenic zones of the Mesozoic and Tertiary (summary by Dickinson, 1970). Engel et al. (1974), however, have proposed that Proterozoic and Paleozoic orogenic belts are

dominantly ensialic; they base this conclusion on the small proportions of graywackes and related rocks of low K_2O/Na_2O ratio
formed at these times in comparison with arkoses, quartzites,
and other, more sialic, materials.

It is not clear whether the observations of Engel et al.
(1974) indicate any variation with time in the absolute amount
of graywacke being produced in the Earth. Possibly the variation
in ratios that they propose is not caused by any decrease in the
amount of graywacke production but simply by an increase in the
amounts of granitic, arkosic debris in the Proterozoic and Paleozoic. This possibility is supported by the appearance, mentioned
above, of significant volumes of graywackes of all ages.

Assuming that considerable quantities of graywacke have
been formed during all periods of geologic time, the next question to be considered is the nature of their source terranes.
Graywackes consist largely of quartz, volcanic rock fragments,
plagioclase, and a clay matrix. The volcanic debris clearly is
derived from the volcanic sequences that commonly underlie the
graywackes and associated sediments, and the quartz and plagioclase are obtained from plutonic rocks.

The plutonic source rocks for the Archean (Yellowknife)
graywackes in the Slave Province of the Canadian shield have
been considered by Henderson (1972) to be the typical sialic complex of the crystalline shield. This conclusion is based on
the occurrence of rare granite fragments in the graywackes,
overlap of sediment onto eroded crystalline basement, evidence
of sediment transport from the present margins into the basin,
and a similarity of composition of the graywackes to the composition of the exposed crystalline rocks of the present Canadian shield. The compositional similarity of graywackes and
the crystalline rocks of shields is not restricted to the
Yellowknife area; Pettijohn et al. (1972) discuss the similarity
of graywacke and granodiorite compositions, and the work of Eade
and Fahrig (1971) has shown that the Canadian shield has essentially the composition of average granodiorite. Condie (1967)
uses similar relationships in some Wyoming graywackes to infer
a granodioritic composition for Archean crust.

The concept that sialic, plutonic source rocks are major
contributors to Archean graywackes seems to be incompatible
with the conclusion discussed by Rogers and McKay (1972) that
graywacke is formed by the erosion of primitive, mantle-derived
material. Rogers and McKay considered the source rocks of
graywackes to be oceanic and island arc volcanic rocks plus
the types of plutons (mainly quartz diorites, with a very low
K_2O/Na_2O ratio) that typically occur in island arcs and other
noncratonic environments. It is possible, however, that the
crystalline basement existing at the time of formation of Archean
greenstone belts is compositionally more similar to the plutonic
rocks of island arcs than to the present composition of exposed

Precambrian crust.

The possibility that crystalline source rocks of greenstone belts and related Archean graywacke environments are similar in composition to the quartz diorite of very low K_2O/Na_2O
ratios in islands arcs is supported by two observations:

(1) Graywackes commonly contain even less potassium feldspar than the average granodiorite. The average granodiorite
of Nockolds (1954), which is commonly cited for comparison purposes, contains 18.3% normative orthclase (although probably
less modal potassium feldspar). Henderson (1972) indicates no
potassium feldspar in the Yellowknife graywackes, and other workers confirm its scarcity or absence in most graywackes (reviews
by Pettijohn et al., 1972; Rogers and McKay, 1972). Henderson
explains the total absence of potassium feldspar by diagenetic
or postdepositional destruction. Whereas destruction within
the sediment may be important, it is also possible to interpret
the extremely low concentration of potassium feldspar as being
indicative of the erosion of rocks more mafic than average
granodiorite.

(2) The average granodioritic composition of shields in-
corporates the composition of rocks formed both before and after
the Archean greenstone belts. The geochronologic relationships
between basement, greenstones, and later intrusions are commonly
very difficult to decipher, but most general studies clearly
indicate that younger intrusions tend to be more potassic than
older ones (e.g., Hunter, 1974a,b). Thus, the composition of
the 3000- to 3500-million year old rocks that presumably composed the basement source for most Archean greenstones may be
more mafic than the average of the present exposed crust, with
the present composition controlled partly by intrusion and met-
asomatism after greenstone formation. Krupicka (1975) discusses the problem of the amount of true granite in the earliest
Archean crusts.

The studies of shield evolution mentioned earlier may lend
some support to the concept of a relatively mafic early Archean
crust, particularly one with a very low K_2O/Na_2O ratio. Table
3 shows the K_2O and Na_2O contents of the Archean and Proterozoic
crusts of Canada (from Eade and Fahrig, 1971) and the Llano uplift modified from Johnson (1975). Assuming that the Archean
crust stabilized at 2600 million years ago, the Proterozoic
(essentially the Churchill Province) at 1800 million years, and
the Llano at 1000 million years, it is possible to make a rough
linear extrapolation back to the composition of a crust stabilized
3500 million years ago. This postulated 3500-million-year-old
crust should have a K_2O content of 1.5% and a Na_2O content of 5.0%.

A plutonic rock with only 1.5% K_2O probably has most of its
potassium incorporated in biotite and little or none in potassium
feldspar. Thus, sediment derived from these very ancient crus-

tal rocks would have virtually no potassium feldspar and would be comparable to the sediment formed in younger orogenic belts.

TABLE 3

K_2O and Na_2O in Continental Crust of Different Ages

	Age (million years)	K_2O %	Na_2O %
Llano[a]	1000	4.4	3.3
Canadian Proterozoic[b]	1800	3.5	3.5
Canadian Archean[b]	2500	2.7	4.1
Early, pregreenstone, Archean[c]	3500	᠕1.5	᠕5.0

[a]Llano modified from Johnson (1975).

[b]Canadian Proterozoic and Archean from Eade and Fahrig 1971.

[c]Early, pregreenstone, Archean estimated by linear extrapolation of K_2O - age and Na_2O - age relationships.

The point of this discussion is that the composition of graywackes of all ages indicates that at least some, if not all, of their source rocks were relatively primitive material recently derived from the mantle. This observation indicates incorporation of mantle-derived material into the continental craton throughout geologic time.

ACKNOWLEDGMENTS

I would like to thank Drs. Francis Pettijohn and John Henderson for their thoughtful discussions of some of the problems discussed in this chapter.

REFERENCES

Anhaeusser, C. R. 1973. In J. Sutton and B. F. Windley (eds.), *A Discussion on the Evolution of the Precambrian Crust,* Phil. Trans. Roy. Soc. London, ser. A, Vol. 273, pp. 359 - 388.

Armstrong, R. L., and Hein, S. M. 1973. *Geochim. Cosmochim. Acta,* 37:1.

Ben-Avraham, Z., and Emery, K. O. 1973. *Amer. Assoc. Petrol. Geol. Bull.,* 57:2323.

Card, K. D., Church, W. R., Franklin, J. M., Frarey, M. J., Robertson, J. A.,, West, G. F., and Young, G. M. 1972.

In R. A. Price and R. J. W. Douglas (eds.), *Variations in Tectonic Styles in Canada,* pp. 335 - 380. Geological Association of Canada, Toronto, Canada.

Condie, K. C. 1967. *Science, 155*:1013.

Condie, K. C., Macke, J. E., and Reimer, T. O. 1970. *Geol. Soc. Amer. Bull., 81*:2759.

Dickinson, W. R. 1970. *Rev. Geophys. Space Phys., 8*:813.

Donaldson, J. A., McGlynn, J. C., Irving, E., and Park, J. K. 1973. In D. H. Tarling and S. K. Runcorn (eds.), *Implications of Continental Drift to the Earth Sciences,* Vol. 1, pp. 3 - 17. Academic Press, New York.

Donnelly, T. W., Rogers, J. J. W., Pushkar, P., and Armstrong, R. L. 1971. In T. W. Donnelly (ed.), Caribbean Geophysical, Tectonic, and Petrologic Studies, *Vol. 130, pp. 181 - 244. Geol. Soc. Amer. Mem.*

Eade, K. E., and Fahrig, W. F. 1971. *Geol. Surv. Canada Bull., 179*:51.

El Shazly, E. M., Hashad, A. H., Sayyah, T. A., and Bassyuni, F. A. 1973. *Egypt J. Geol. 17*:1.

Engel, A. E. J., Itson, S. I., Engel, C. G., Stickney, D. M., and Cray, E. J., Jr. 1974. *Geol. Soc. Amer. Bull., 85*:843.

Fleck, R. J. 1972. *EOS* (abstr.), *53*:1130.

Glikson, A. Y., and Lambert, I. B. 1973. *Earth Planet. Sci. Lett. 20*:395.

Hashad, A. H., Sayyah, T. A., ElKholy, S. B., and Youssef, A. 1972. *Egypt. J. Geol., 16*:269.

Helmbold, R., and Van Houten, F. B. (translator) 1958. *Geol. Soc. Amer. Bull., 69*: 301.

Henderson, J. B. 1972. *Canad. J. Earth Sci., 9*:882.

Henderson, J. B. 1975. *Canad. J. Earth Sci., 12*:1619.

Hunter, D. R. 1974a. *Precambrian Res., 1*:259.

Hunter, D. R. 1974b. *Precambrian Res., 1*:295.

Hurley, P. M. 1972. *Earth Planet. Sci. Lett. 15*:305.

Hurley, P. M. 1974. *Geology, 2*:373.

Jakes, P., and White, A. J. R. 1972. *Geol. Soc. Amer. Bull., 83*:29.

Johnson, L. A. 1975. General chemical composition of Precambrian crust in the Llano uplift, central Texas. Master's thesis, Rice University, Houston, Texas.

Katili, J. A. 1975. *Tectonophysics, 26*:165.

Krupicka, J. 1975. *Canad. J. Earth Sci., 12*:1307.

McGlynn, J. C., and Henderson, J. B. 1972. In R. A. Price and R. J. W. Douglas (eds.), *Variations in Tectonic Styles in Canada,* pp. 505 - 526. Geological Association of Canada,

Mitchell, A. H. G. 1975. *Geol. Soc. Amer. Bull., 86*; 1487.

Muehlberger, W. R., Denison, R. E., and Lidiak, E. G. 1967. *Amer. Assoc. Petrol. Geol. Bull., 51*; 2351.

Nockolds, S. R. 1954. *Geol. Soc. Amer. Bull., 65*; 1007.

Pettijohn, F. J. 1970. In A.J. Baer (ed.), *Symposium on Basins and Geosynclines of the Canadian Shield,* pp. 239 - 255. Canada Geol. Surv. Paper 70 - 40.

Pettijohn, F. J. 1972. In B. R. Doe and D. K. Smith (eds.), *Studies in Mineralogy and Precambrian Geology,* pp. 131 - 149. Geol. Soc. Amer. Mem. 135.

Pettijohn, F. J., Potter, P. E., and Siever, R. 1972. *Sand and Sandstone.* Springer-Verlag, Berlin.

Rogers, J. J. W., Burchfiel, B. C.,Abbott, E. W., Anepohl, J. K., Ewing, A. H., Koehnken, P. J., Novitsky-Evans, J. M., and Talukdar, S. C. 1974. *Geol. Soc. Amer. Bull., 85:*1913.

Rogers, J. J. W., and McKay, S. M. 1972. In B. R. Doe and D. K. Smith (eds.), *Studies in Mineralogy and Precambrian Geology,* pp. 3 - 28. Geol. Soc. Amer. Mem 135.

Said, R. (ed.). 1972. *Geol. Surv. Egypt Ann. 2:*281.

Stockwell, C. H. 1968. *Canad. J. Earth Sci., 5:*693.

Wise, D. U. 1974. In C. A. Burk and C. L. Drake (eds.), *The Geology of Continental Margins,* pp. 45 - 58. Springer-Verlag, Berlin.

Wynne-Edwards, H. R. 1972. *In R. A. Price and R. J. W. Douglas (eds.), Variations in Tectonic Styles in Canada,* pp. 263 - 334.. Geological Association of Canada, Toronto, Canada.

Zartman, R. E. 1964. *J. Petrol., 5:*359.

Zartman, R. E. 1965. *J. Petrol., 6:*28.

5

ARCHEAN GEOLOGY AND EVIDENCE OF ANCIENT LIFE IN THE SLAVE STRUCTURAL PROVINCE, CANADA

JOHN B. HENDERSON
Geological Survey of Canada

The Archean supracrustal rocks in the Slave Structural Province, Northwest Territories, Canada, consist of conformable sequences of mainly greywacke-mudstone turbidites of felsic volcanic provenance and of volcanic sequences that range from subaqueous mafic pillow lavas to subaerial felsic volcanic deposits. No sediments are known that indicate deposition under geologically stable conditions during this time. Preserved remnants of Archean basins consist of turbidite-filled basins with marginal volcanic buildups. All shallow water or subaerial sedimantary deposits that are known occur at these basin margins. The basins are separated by extensive, and in some cases, very complex granitic terrains that typically have intrusive relationshops with the supracrustal rocks. Basement to the supracrustal successions, where identified or inferred, is everywhere granitic in composition. The time represented by the supracrustal succession at a given locality in the province is relatively short, possibly on the order of 15 million years. Evidence of life in the Archean of the Slave Structural Province is present but sparse. In places associated with the turbidites, carbonates, and certain volcanic environments there are units of black carbonaceous shales. This carbon may be of biogenic origin. More convincing evidence of life, however, is the occurrence of stromatolites in a volcanic-sedimentary sequence in the northern part of the province.

The Canadian Shield has been divided into several structural provinces based on internal structural trends imposed at various times during the Precambrian (Stockwell, 1970). The

Slave and larger Superior Structural Provinces have remained relatively stable since the end of the Archean[1] 2.5 billion years ago, while the intervening Churchill Structural Province, even though made up of a high proportion of Archean rocks, continued to undergo orogenic events until the middle of the Proterozoic Eon. Thus, the Slave Province should not be thought of as an isolated "island" of Archean in a sea of younger rocks, but rather as a well-preserved remnant of a much larger Archean "continent," major parts of which were caught up in later, post-Archean, orgenic events.

The Slave Province is located in the northwestern part of the Canadian Shield, between Great Slave Lake and Coronation Gulf, where it occupies an area of about 190,000 square kilometers (Fig. 1). Its borders with adjacent younger provinces are marked by unconformities or structural discordances. The province is underlain by granitic rocks and by supracrustal rocks that are typically highly deformed and variously metamorphosed. The granitic rocks underlie approximately half the province and for the most part are intrusive into the older sedimentary and volcanic rocks of the Archean Yellowknife Supergroup. The Yellowknife supracrustal succession, as is commonly the case in most Archean terrains of the world, is simple, consisting primarily of thick sequences of predominantly mafic subaqueous volcanic rocks and greywacke-mudstone sediments. The Yellowknife Supragroup differs from most Archean supracrustal successions in that sedimentary rocks greatly exceed the volcanic rocks in volume.

Due to the extensive period of granitic intrusion and structural deformation that marked the end of the Archean in this area, the preserved supracrustal rocks are for the most part highly deformed and metamorphosed. It is unusual indeed to see structural attitudes of less than 45°. The volcanic sequences occur in steeply dipping homoclinal successions, while the less competent sediments have been thrown into very complex fold patterns. Although large areas of the province have undergone only low-grade greenschist metamorphism, the metamorphic grade is commonly of the amphibolite facies (Fig. 1). Like most Archean terrains, the metamorphism took place under relatively low pressure conditions.

The Yellowknife sediments show very little variation throughout the province. They consist almost exclusively of greywackemudstone turbidites with very minor amounts of other sedimentary rocks, including some carbonates, fluvial sandstones,

---

[1]The term Archean is used here in the sense of Stockwell (1970): "The Archean Eon includes rocks involved in or emplaced during the Kenoran Orogeny and all older rocks."

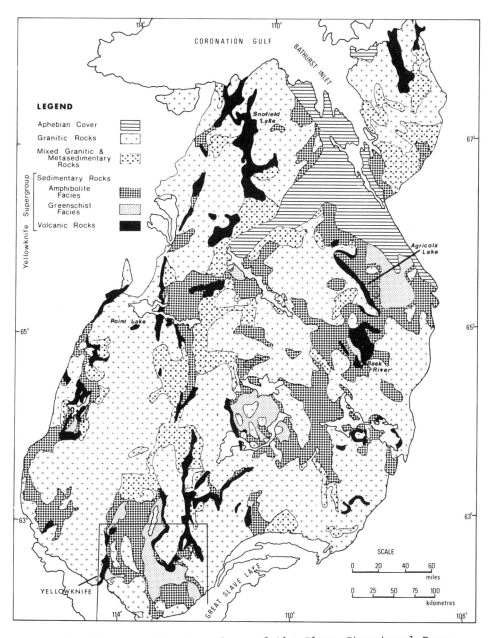

Fig. 1 Generalized geology of the Slave Structural Province. Locations cited in the text are noted. Area outlined
in the southern part of the province east of Yellowknife is
the area represented in Fig. 2.

conglomerates, and iron formations of various types. Sediments
reflecting deposition under geologically stable conditions so
common in the post-Archean Precambrian are unknown. The Yellowknife volcanic rocks are mainly of basaltic to andesitic composition, although there is a higher proportion of felsic volcanic
rocks than has been reported in many other Archean areas. This
is particularly true of certain belts that are composed almost
entirely of felsic material, as in the Back River area (Fig. 1)
and some belts in the northern part of the structural province.

The preserved areas of supracrustal rocks are interpreted
as remnants of distinct depositional basins that were separated
by positive areas now represented by vast areas of granitic
rocks (McGlynn and Henderson, 1972). Three such basins in the
structural province have been proposed and include the narrow
zone along the western margin, a northerly trending zone in the
central part that has been disrupted in the center by extensive
granitic intrusion and migmatization of the supracrustal rocks
and the large equidimensional area in the eastern part (Fig. 1).
One of the best preserved and exposed parts of one of these
basins occurs in the southern part of the province east of the
city of Yellowknife. This area will be discussed as an example
of the geology representative of the Slave Structural Province
as a whole.

I.   YELLOWKNIFE-HEARNE LAKE AREA

The basin segment contained in this area (Fig. 2) is bordered in part by linear volcanic belts, while the central part
of the basin is occupied by greywacke-mudstone turbidite deposits.

The typical volcanic sequence consists of more or less
equal proportions of massive and pillowed mafic flows (Henderson and Brown, 1966; Baragar, 1966). The volcanic piles are
thick, with the sequence at Yellowknife on the west side of the
basin at least 7000 m thick. Individual flows vary in thickness from less than a meter to as much as 150 m and are equally
variable in exposed lateral extent. Pillow structures within
the flows indicate subaqueous extrusion. The flows are extensively intruded by dykes, sills, and irregular intrusions of
similar composition to the flows and are probably more or less
contemporaneous with volcanism. Sedimentary units are rare,
although minor tuffaceous units occur in the dominantly mafic
volcanic sequences. Although small amounts of intermediate and
felsic volcanics occur in many of the mafic sequences, they form
the major part of the volcanic pile west of Desperation Lake
(Fig. 2). Major buildups of felsic volcanics also occur away
from the basin margin within the turbidite sequence, as seen
north of Yellowknife and west of Duncan Lake. The presence of

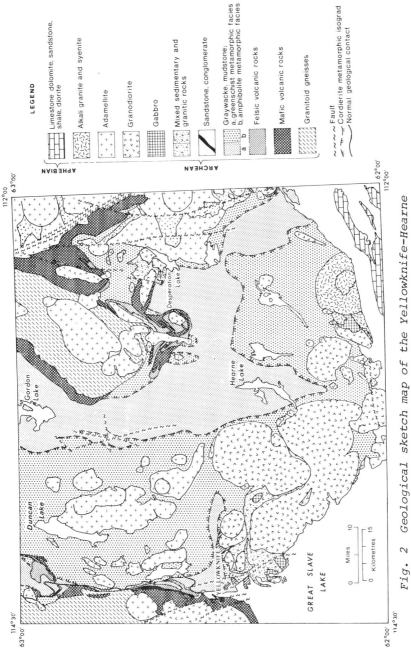

Fig. 2 Geological sketch map of the Yellowknife–Hearne Lake area in the southern Slave Province.

thick welded ash-flow tuff units suggests that at least some
of the felsic volcanic accumulations were emergent islands
(Lambert, 1974).

The sedimentary basinal deposits are dominantly a homogeneous sequence of greywacke-mudstone turbidites indicative of
subaqueous deposition, probably in deep water, and certainly
below wave base. East of Yellowknife the turbidite sequence
is at least 5000 m thick. Primary sedimentary structures
characteristic of turbidity current deposition, such as graded
bedding, and various soft-sediment deformation structures, are
commonly well preserved, even in some of the more highly metamorphosed rocks. Detailed study of the turbidites at Yellowknife (Henderson, 1972) indicates that the sediments were deposited on a complex of submarine fans near the basin edge. On
this fan, complex thick beds of coarser-grained material were
deposited in depositional fan valleys with thinner beds of finer-
grained sediments accumulating on the interfan valley areas of
the fan. Paleocurrent measurements in this area indicate that
sediment transport was from the west, approximately perpendicu-
lar to the present exposed boundary of the basin. The turbidite
sandstones are mineralogically, chemically, and texturally
very immature, and thus are a reasonably unaltered sample of
the terrain from which they were derived. The main components
are quartz, rock fragments, and feldspar. It appears that these
rock fragments and minerals were derived almost exclusively
from a source composed of felsic volcanics and granitic intru-
sives. There is no evidence of older sediments or their metamorphosed equivalents, and mafic volcanic material forms only a
very minor part of the source area. The provenance of the sediments has important implications as to the nature of the crust
at this time.

Minor amounts of sediments other than turbidites also o-
ccur in the region. These occur only at the basin margins.
Carbonates, including both dolomite and limestone, occur with
the felsic volcanic units, commonly at the transition between
felsic volcanism and "normal" basinal sedimentation. In the
volcanic complex west and north of Desperation Lake (Fig. 2)
such a carbonate unit can be discontinuously traced for a distance of 50 km. Black, commonly siliceous, and sulphide-bear-
ing carbonaceous mudstones are associated with the carbonates.
Zinc - copper sulfide mineralization, where present, is also
commonly found at this stratigraphic position. Fluvial sediments are very rare and are presently known from only two localities in the province, one of which is at Yellowknife.
These sediments lie unconformably on the mafic volcanic pile
and represent a braided river environment. Like the turbidites,
they had a dominantly felsic volcanic source and represent the
subaerial basin margin equivalent of the deep water basinal

turbidites. Conglomerates, where present, also only occur at
the basin margins. At Yellowknife, they fill depressions in
the above-mentioned unconformity surface with the clasts derived mainly from the underlying mafic volcanics with a minor
granitic boulder component.

The Yellowknife supracrustal rocks are now contained and
intruded by a variety of granitic rocks. The youngest intru-
sives are massive, generally more potassic, adamellites that
occur along the crest of a broad thermal ridge (as indicated
by the metamorphic isograds in Fig. 2) in the central part of
the basin and also along the eastern side. These plutonic complexes are strongly discordant to the structure in the supracrustal rocks they intrude. Inclusions of metasediments in
the granitic rocks are commonly undeformed and the plutons
themselves are not deformed. Somewhat older, large, massive
plutonic complexes of granodiorite also intrude the supracrustal
succession. Individual plutonic lobes are outlined by screens
of typically highly deformed to gneissic inclusions of metasediments. Compared to the adamellites, the granodiorite plutons
tend to be more concordant to the structure in the surrounding
country rocks. They are characterized by a very narrow zone
of higher-grade metamorphism in the adjacent metasediments. The
third major group of granitic rocks occurs west of Duncan Lake
and between Gordon Lake and Desperation Lake. This group is
different in that it is strongly deformed, metamorphosed, and
quite heterogeneous, contrasting strongly with the massive generally homogeneous plutons of the other two units. Compositions
range from metagabbro through to granodiorite, with an "average"
probably approximating that of a quartz diorite.

The second group of granitic rocks, the granodiorites, are
of some interest as the author feels these plutonic complexes
may be related to the source of much of the sedimentary material in the basin. The sediments were derived from a mixed felsic volcanic and granitic terrain. Chemical analyses of the
sediments and of an average granodiorite show certain similarities (Table 1). It is suggested that the granodiorite plutons
may be comagmatic but later-emplaced deeper parts of epizonal
plutons originally intruded into their own volcanic debris.
These felsic volcanics and comagmatic high level plutons provided detritus that accumulated in the adjacent sedimentary basins.

It has been suggested that the third group of granitic rocks,
the heterogenous granitoid gneisses, are older than the Yellowknife supracrustal rocks (Baragar, 1966; Davidson, 1972). East
of Gordon Lake (Fig. 2) there is a strong structural and metamorphic contrast between the granitoid gneisses and the overlying volcanic rocks. In this region, basaltic dykes interpreted as feeders to the overlying volcanic flows intrude only

TABLE 1

	1a[a]	*1b*[a]	*2b*	*3c*	*4d*	*5e*
				Chemical analyses		
SiO_2	66.24	53.26	66.88	65.2	49.21	60.74
Al_2O_3	15.28	20.64	15.66	15.8	15.81	15.83
Fe_2O_3	0.70	1.26	1.33	1.2	2.21	0.50
FeO	4.53	7.13	2.59	3.4	7.19	1.80
MgO	2.74	4.81	1.57	2.2	8.53	2.04
CaO	1.70	1.24	3.56	3.3	11.14	4.03
Na_2O	3.12	2.20	3.84	3.7	2.71	0.39
K_2O	1.91	3.53	3.07	3.23	0.26	3.95
H_2O^+	2.49	4.62	0.65	–	–	1.45
H_2O^-	0.08	0.11	–	0.08	–	0.11
CO_2	0.38	0.10	–	0.20	–	5.55
TiO_2	0.64	0.93	0.57	0.57	1.39	0.67
P_2O_5	0.12	0.16	0.21	0.17	0.15	0.25
MnO	0.06	0.09	0.07	0.08	0.16	0.08
S	–	–	–	–	–	1.45
Cl	–	–	–	–	–	0.03
Cr_2O_3	–	–	–	–	–	0.01
C	–	–	–	–	–	1.28
Other	–	–	–	0.11	–	–
Total	99.99	100.08	100.0	100.0		100.16

[a]Average of three composite samples of greywacke (1a) and mudstone (1b) at Yellowknife (Henderson, 1972).

[b]Average granodiorite (Nockolds, 1954).

[c]Average composition of the Canadian Shield (Eade and Fahrig, 1971).

[d]Average oceanic basalt (Melson, 1968).

[e]Composite sample of graphitic tuff, Yellowknife (Boyle, 1961).

the granitoid rocks. Thus, it appears that these granitoid gneisses may be basement to the Yellowknife supracrustal rocks and a sample of the early Archean crust in the province.

II.   NATURE OF THE ARCHEAN CRUST IN THE SLAVE PROVINCE

One of the greatest areas of disagreement among workers in Archean rocks has to do with the nature of the Archean crust. It is evident that by the end of the Archean, after the major period of intrusion and deformation that marks the end of the Archean, there was a high proportion of granitic rocks in at least those parts of the Archean crust that have been preserved (Fig. 1). Less apparent is the nature of the crust on which the Archean sediments and volcanic rocks were deposited. One school of thought holds that the Archean supracrustal rocks were deposited on oceanic or simatic crust (Glickson, 1972; Arth and Hansen, 1975; Folinsbee *et al.*, 1968), while others believe that the supracrustal rocks were ensialic-- underlain by granitic continental crust (Windley and Bridgwater, 1971; Hunter, 1974; McGlynn and Henderson, 1970).

Evidence of the nature of the crust from the Slave Province supports the "granitic school," in the opinion of the author. Chemical analyses of the sediments indicate that they are chemically more similar in composition to a granodiorite (representative of sialic crust) than a basalt (representative of simatic crust) (Table 1). Of some interest in this regard is the similarity of the chemical composition of the Canadian Shield of today (Table 1). This corresponds with the petrographic evidence which indicates that the source of the sediments consisted dominantly of a mixed terrain of *felsic* volcanics and granitic rocks. When the volume of sediments of this composition in the province is considered, the source area for this amount of sialic material must have been very large indeed. The existence of an extensive sialic crust is further supported by gravity measurements throughout the province (Hornal and Boyd, 1972) as there is no indication that the sedimentary basins are underlain by a significant thickness of mafic material that might represent Archean oceanic crust.

More direct evidence of sialic basement is the occurrence in several locations of granitic rock clearly older than the supracrustal successions. One such example already mentioned is in the Yellowknife-Hearne Lake area east of Gordon Lake (Fig. 2). Others include the occurrence of granitic boulders presumably derived from the underlying basement rocks in a diatreme that intrudes the volcanic rocks at Yellowknife. These boulders have been dated at 3030 million years (Nikic *et al.*, 1975), while the volcanic rocks have been dated at about 2650 million years (Green and Baadsgaard, 1971). In the western part of the province, granitic gneisses have been dated at about 3.0 billion years (Frith *et al.*, 1974). Perhaps the best evidence for the existence of granitic basement to the supracrustal rocks was recognized by Stockwell (1933), over 40 years ago. At Point Lake (Fig. 1), some 300 km north of Yellowknife, at a similar basin margin, an unconformity is clearly exposed with conglomerate and fluvial sediments similar to those at Yellowknife, clearly overlying the granitic basement (Fig. 3) (Henderson, 1975a). Preliminary dating of this granitic basement indicates an age of 3.0 billion years (R. K. Wanless, personal communica-

Fig. 3 Unconformity at Point Lake between granitic base-ment (lighter color) and the overlying conglomerate consisting mainly of mafic volcanic clasts with minor granitic and felsic cobbles. Note fractures in the unconformity surface containing cobbles of the conglomerate. Scale is 43 cm long.

tion, 1975).

These granitic basement dates clustering at about 3 bill-ion years contrast strongly with available estimates of time of sedimentation and volcanism which group at about 2.7 billion years.

III. THE EXTENT OF THE ARCHEAN RECORD IN THE SLAVE PROVINCE

The Archean geology throughout the Slave Province is essen-tially similar to that described above for the basinal segment east of Yellowknife, but it is not known if sedimentation and volcanism everywhere in the province were contemporaneous. Assuming that certain radiometric dates can be interpreted as the age of volcanism or sedimentation, the preliminary, very incomplete, and limited data available indicate that the accum-ulation of the supracrustal rocks may have been contemporaneous with deposition about 2.7 billion years ago (Green and Baads-gaard, 1971; R. K. Wanless, personal communication, 1975). This, then, presents a rather curious problem. The time repr-sented by the Archean record as preserved at any given locali-ty is very short. For example, Folinsbee et al. (1968) sugges-

ted that the supracrustal rocks at Yellowknife probably accumulated in about 15 million years. They pointed out the similarity in character and thickness of the sequence at Yellowknife to the Miocene sediments and volcanic rocks of the Fossa
Magna of Japan. From paleontological evidence the Miocene example is known to have accumulated in about 15 million years
and so by analogy it is reasonable to assume that the sequence
at Yellowknife could have been deposited in a similar length
of time. No *major* unconformities representing extended periods
of time are known to separate sequences of supracrustal rocks;
the only time break known being the gap between the granitic
basement and the start of the brief supracrustal record. Thus,
we are left with the anomaly of the Archean Eon which includes
nearly half of geologic time represented by a depositional record of about 15 million years. Between the age of the known
basement rocks and the end of the Archean is a time period of
about 500 million years--roughly similar to the length of the
Phanerozoic Eon. The preserved depositional record represents
only about 5% of this time. Since the record is so short, the
Archean supracrustal rocks of the Slave Province cannot really
be considered as representative of conditions during this major period of Earth history. Our knowledge of this time period
in this area will always be at best, very incomplete.

IV. EVIDENCE OF ANCIENT LIFE IN THE ARCHEAN OF THE SLAVE PRO-
     VINCE

     Evidence of ancient life in the Archean of the Slave Pro-
vince is present, but sparse, due to the metamorphic and tectonic disruption of the rock, the easily degraded nature of
such material, and the fact that much of the province has only
been explored on a reconnaissance scale.
     One possible indicator of the existence of biologic acti-
vity is the carbon-rich sedimentary deposits which are present
throughout the province. They occur as massive, rather feature-
less, beds of black mudstone on the order of a meter in thickness. They are quite distinctive from the pelitic mudstones
normally present with the turbidite greywackes and siltstones
with which they are associated. The greater thickness of these
carbonaceous beds compared to the thickness of the normal mudstone between the greywacke layers indicates a much larger period of time than is normally believed to occur between success-
ive turbidity currents. Some thin tuff beds containing a high
proportion of sulfur and carbon occur within the volcanic sequence at Yellowknife. The carbon in these units (Table 1)
has been suggested as being biogenic (Boyle, 1961). Carbonates,
where present, are commonly associated with similar black carbonaceous mudstones. For example, in the Back River area (Fig.
1), at the margin between a large, dominantly felsic volcanic

pile and a flanking area of mixed sediments and volcanics, is
a zone of sediments dominated by carbonaceous mudstone and
chert - carbonate iron formation (Henderson, 1975a). The massive black carbonaceous mudstone commonly contains finely disseminated pyrite to spectacular pyrite concretions several centi-
meters in diameter. Similarly, in the Agricola Lake area (Fig.
1), an elongate, 200-m-thick lens of black, severely deformed
carboraceous mudstones, locally containing finely disseminated
to concretionary pyrite, occurs between intermediate to felsic
volcanics and the normal greywacke-mudstone turbidites. In the
volcanics below the thickest part of the lens is a base metal
(Zn, Cu) mineral deposit thought to be more or less contemporaneous with volcanism. The association of black carbonaceous
mudstones, carbonates, felsic volcanics, and base metal mineralization is one that is seen in many parts of the Slave Province (Henderson, 1975b). If the carbonaceous mudstones are of
biogenic origin, the base metal mineralization processes may
have had some influence on their formation. The volcanic exhal-
ations that presumably were involved with the formation of the
mineral deposits may have acted as a source of nutrients for
the life forms and so encouraged their local proliferation. On
the other hand, or perhaps at the same time, they may have
formed local toxic environments that are now represented in the
record as carbon-rich deposits.

The most convincing evidence for the existence of life
during the Archean is the occurrence of well-preserved stromatolites near Snofield Lake in the northern part of the province
(Fig. 1) (Henderson, 1975b). Stromatolites are biogenic structures commonly found in carbonate rocks that form due to the
sediment trapping and binding capabilities and/or carbonate-
precipitating abilities of blue-green algae, and to a lesser
extent, bacteria. Arguments as to the biogenicity of these
Archean forms are based only on the similarity of the Archean
structures with more recent forms of known biogenic origin. No
remnants of algal filaments or cells have been recognized. As
with carbonates elsewhere in the province, the stromatolites
occur at the contact between a major felsic volcanic buildup
and the normal greywacke-mudstone sediments. The most common
stromatolitic form is a flat variety with wavy to corrugated
laminations that contains small local convexities with relief
of about 1 cm. A few small stromatolitic heads to low mounds
that are formed with similar even to corrugated laminae also
occur (Fig. 4). Minor intraformational breccia layers are present, which in some cases contain clasts that have formed onco-
lites. Although well-preserved stromatolites are presently
known from only the one locality in the Slave Province, similar
more deformed and more highly metamorphosed carbonate units
occur at similar stratigraphic positions, presumably repre-

Fig. 4 Archean stromatolite consisting of gently curved wavy to corrugated laminations of presumed biogenic origin.

senting similar environments throughout the province. This suggests that stromatolites may have been more abundant and extensive than is presently indicated.

Acknowledgments

The manuscript was read by F.H.A. Campbell, J.C. McGlynn, and M. Schau of the Geological Survey of Canada, whose suggestions for improvement are greatly appreciated.

REFERENCES

Arth, J. G., and Hansen, G. N. 1975. *Geochim. Cosmochim. Acta. 39*:325.

Baragar, W. R. A. 1966. *Canad. J. Earth Sci., 3*:9.

Boyle, R. W. 1961. *Geol. Surv. Canad. Mem., 310.*

Davidson, A. 1972. *Geol. Surv. Canad. Pap. 72-1A,* p. 109.

Eade, K. E., and Fahrig, W. F. 1971. *Geol. Surv. Canad. Bull., 179.*

Folinsbee, R. E., Baadsgaard, H., Cumming, G. L., and Green, D. C. 1968. *Amer. Geophys. Union, 12*:441.

Frith, R. A., Frith, R., Helmstaedt, H., Hill, J. and Leatherbarrow, R. 1974. *Geol. Surv. Canad. Pap. 74-1A,* p. 165.

Glickson, A. Y. 1972. *Geol. Soc. Amer. Bull., 83*:3323.

Green, D. C., and Baadsgaard, H. 1971. *J. Petrol.* *12*:177.

Henderson, J. B. 1972. *Canad. J. Earth Sci.*, *9*:882.

Henderson, J. B. 1975a, *Geol. Surv. Canad. Pap. 74-1A,* p. 325.

Henderson, J. B. 1975b. *Canad. J. Earth Sci.*, *12*:1619.

Henderson, J. F., and Brown, I. C. 1966. *Geol. Surv. Canad. Bull.*, *141*.

Hornal, R. W., and Boyd, J. B. 1972. *Dept. Energy, Mines, and Resources, Canada, Earth Physics Branch, Gravity Map Series,* Nos. 89-95.

Hunter, D. R. 1974. *Univ. Witwatersrand Econ. Geol. Research Unit Inf. Circ.*, 83.

Lambert, M. B. **1974.** *Geol. Surv. Canad. Pap. 74-1A,* p. 177.

McGlynn, J. C., and Henderson, J. B. 1970. *eol. Surv. Canad Pap. 70-40,* p. 31.

McGlynn, J. C., and Henderson, J. B. 1972. *Geol. Assoc. Canad. Spec. Pap.*, *11*:506.

Melson, W. G., Thompson, G., and Van Andel, T. H. 1968. *J. Geophys. Res.*, *73*:5925.

Nikic, A., Baadsgaard, H., Folinsbee, R. E., and Leech, A. P. 1975. *Geol. Soc. Amer. Abstr. Prog.* 7:1213.

Nockolds, S. R. 1954. *Geol. Soc. Amer. Bull.* 65:1007.

Stockwell, C. H. 1933.*Geol. Surv. Canad. Ann Rep. 1932, pt. C,* p. 37.

Stockwell, C. H. 1970. *Geol. Surv. Canad. Econ. Geol. Rep. 1*:43

Windley, B. F., and Bridgwater, D. 1971. *Spec, Publ. Geol. Soc. Austral. 3*:33.

6

CARBON CONTENTS OF EARLY PRECAMBRIAN ROCKS

CARLETON B. MOORE and DONNA WELCH
Arizona State University, Tempe

The question of whether there are detectable chemical parameters in Early Precambrian sediments or metasediments that will
shed light on geochemical conditions early in the Earth's evolutionary history is an interesting and critical one. The
paucity of sedimentary rocks of great age makes the detection
and evaluation of critical chemical components difficult. The
element carbon in its reduced or organic state is a critical
element in attempting to evaluate the time of beginning and
origin of biogenic compounds.

Studies by Ronov and co-workers (Ronov, 1958) have indicated
a roughly downward trend in the concentration of the average
content of organic carbon in sedimentary rocks with increasing
age. Sedimentary rocks from both the Russian Platform and the
United States show parallel trends with a notable low in Triass-
ic rocks and a high in Carboniferous rocks. Recent work by
Jackson and Moore (1976) and Jackson (1975) on carbon contents
in pre-Phanerozoic and Phanerozoic sedimentary rocks showed
surprisingly regular trends with respect to age. Mudstones
and carbonates showed a progressive increase in organic carbon
content with decreasing age, as reported by earlier investigators, but cherts showed a decrease in organic content with decreasing age. The cherts studied were among the oldest sediments
available and included those from the Theespruit Formation of
the Swaziland Sequence. These particular cherts were reported
to be higher in organic carbon than younger cherts by Moore,
Lewis, and Kvenvolden (1974). Figure 1 illustrates the secular
variations in organic carbon as reported by Jackson and Moore
(1976).

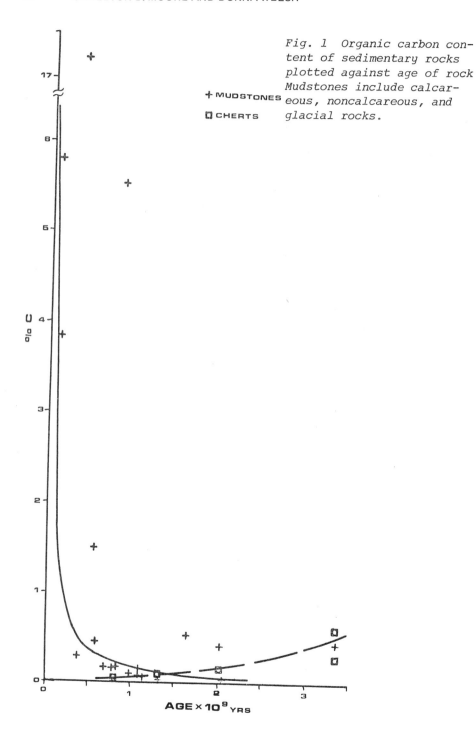

Fig. 1 Organic carbon content of sedimentary rocks
plotted against age of rock
Mudstones include calcar-
eous, noncalcareous, and
glacial rocks.

The majority of the rocks included in the above study was collected by Preston Cloud, as were the samples from the Isua supracrustral belt, West Greenland, discussed below. The discovery of these ancient rocks in Greenland, reportedly about 3.8 million years of age, has allowed us to push back our study of organic carbon even farther in time. It should be noted in passing that organic carbon content is not unique in its regular variation with sample age. Garrels and Mackenzie (1971) have showed regular variations, both positive and negative, in the chemical trends of most analyzed oxides in shaly rocks ranging from Recent to Precambrian in age.

In summary, it appears important to study as many ancient rocks as possible in order to confirm the regular trends in total carbon content observed thus far. In order to remove as many variables as possible from the analyses, great care must be taken to compare rocks of similar petrogenic type. When anomalies or discontinuities in regular trends are noted as in the Swaziland rocks by Moore, Lewis, and Kvenvolden (1974), then inferences may be made of possible changes in chemical evolution and these samples should be subjected to more detailed investigation. Likewise, the chemical nature of the carbon measured should be uniform. Carbonate carbon must be carefully removed by acid treatment before organic carbon is measured. In addition, we have found it useful to remove volatile organic carbon from the total organic carbon by pyrolysis. When this is done we can identify a residual or kerogen carbon component in most rocks. Preliminary results of such studies for a selected suite of rocks are shown in Fig. 2 and Table 1. Note that younger rocks tend to begin to lose volatile carbon at a lower temperature than older rocks and that both seem to reach constant residual carbon contents at 600°C. For this reason we have selected a pyrolysis treatment of 600°C under a flowing nitrogen atmosphere for 2 hr as a suitable treatment to provide for reproducible measure of residual carbon. In Table 1, regularities may be noted in the residual carbon C_R to organic carbon C_0 ratios with age.

Since the Theespruit cherts showed an increase in total carbon content as compared to younger Swaziland Sequence rocks it is of interest to see if the Older Isua supracrustal rocks from Greenland continue this trend.

Table 2 lists some total organic carbon contents of the Isua supracrustal rocks collected by Preston Cloud. The samples are identified by his collection numbers. Specific rock classifications and petrographic descriptions will be published at a later time by other investigators. At the present time the lack of this information diminishes the strength of any conclusion reached on their comparison to general trends. Powdered raw rock samples were leached with dilute HCl and evaporated to dryness before analysis in order to remove carbonate carbon. Noncarbo-

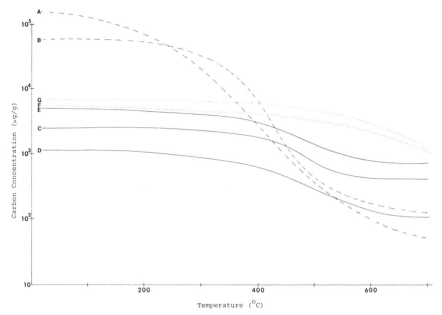

Fig. 2. Pyrolysis temperature profiles for rocks of different ages and thermal histories. A, Green River Shale; B, Pierre Shale; C, Hector Formation, Windermere Shale; D, Nonesuch Shale; E, Fig Tree Shale; F, Isua Gneiss; G, Isua Gneiss.

nate carbon has been considered to organic carbon. The data in Table 2 are published here for the first time. Although some of the rock types included may be of a nonsedimentary history they are included for general information and comparison purposes.

Since cherts appear to possibly contain important data on carbon in ancient rocks, it is useful to compare information on any cherts or metacherts from the Isua supracrustals with other Archean samples. Since the Isua superacrustals have undergone metamorphism, it is somewhat risky to undertake such a comparison both in terms of identifying cherts and in interpreting their carbon abundances. In order to be comparable with the carbon-rich Archean rocks from the Swaziland Sequence, Isua "cherts" should contain organic carbon contents in excess of 0.3 to 0.6%C. The only samples coming close to this range are 1 of 30/8/74 and 2 of 30/8/74B. If, however, these rocks do not have a cherty but a shaly ancestry then a significant number of them contains higher total carbon then would be inferred by extrapolating the mudstone or shale curve of Fig. 1.

It is obvious that the total organic carbon analyses reported and discussed above are quite tentative and speculative, but at this stage they are all we have and they do indicate reg-

TABLE 1 *Age, Temperature Transition, and C_R/C_O Ratio Relationships*

Sample	Age	Transition temperature (°C)	C_O wt %	C_R/C_O[a] (at 600°C)
Green River Shale, Wyoming (A)	40 million years	450	18.6	0.001
Pierre Shale, Wyoming	75 million years	450	5.6	0.002
Hector Formation, Windermere Shale (C)	700 million years	450	0.20	0.3
Nonesuch Shale (D)	1.05 billion years	450	0.11	0.2
Fig Tree Shale, South Africa (E)	3.4 billion years	450	0.31	0.3
Isua Gneiss Greenland (F)	3.8 billion years	600+	0.27	0.8
Isua Gneiss Greenland (G)	3.8 billion years	600+	0.22	0.7

[a] C_R = residual carbon and C_O = total organic carbon.

TABLE 2
Total Carbon Analyses of Isua Greenland Supracrustal Rocks

Sample number	Organic carbon (C_0) (wt %)
1 of 27/8/74	0.009
7 of 27/8/74	0.035
1 of 29/8/74	0.053
1 of 30/8/74	0.48
2 of 30/8/74A	0.022
2 of 30/8/74B	0.21
1 of 31/8/74	0.005
7 of 31/8/74	0.032
3 of 1/9/74	0.070
9 of 1/9/74	0.001
10 of 1/9/74	0.027
1 of 3/9/74	0.002
2 of 3/9/74	0.019

ular trends and interesting discontinuities. The total carbon results indicate that further work on chemistry and isotopic studies is warranted in hopes that critical information on early chemical evolution may become apparent.

Acknowledgments

Preston Cloud collected the majority of the rocks analyzed in this study. Analyses were made in cooperation with Togwell Jackson. This work was supported in part by the Earth Sciences Section National Science Foundation, NSF Grant DES 74-05178.

REFERENCES

Garrels, M., and Mackenzie, F. T. 1971. *Evolution of Sedimentary Rocks*, W. W. Norton, New York.
Jackson, T. A. 1975. *Amer. J. Sci.* 275:906.
Jackson, T. A., and Moore, C. B. 1976. *Chem. Geol*, in press.
Moore, C. B., Lewis, C. F., and Kvenvolden, K. A. 1974. *Precambrian Res.* 1:49.
Ronov, A. B. 1958. *Geochemistry*, 5:510.

7

CONDENSED PHOSPHATES FROM ABIOTIC SYSTEMS IN NATURE

E. J. GRIFFITH

Monsanto Industrial Chemicals Company

The formation of P - O - P linkages in the biosphere is reasonably well understood and many of the reactions have been studied in detail. The formation of condensed phosphates under controlled laboratory conditions is also well known. Conversely, very little is known of the formation of condensed phosphates in the abiotic, uncontrolled systems in nature. There are a number of possible reactions which could yield a small but constantly replenished source of condensed phosphates of nonbiological origin. Although there is little, if any, tangible evidence that these reactions occurred on the Precambrian Earth, it is likely that one or more nonenzymatic routes to condensed phosphates has contributed P - O - P linkages throughout geological time.

I. INTRODUCTION

The Earth's reservoir of condensed phosphates is exceedingly small. It is doubtful that as much as 0.001% of the phosphorus is present in the form of condensed phosphates. Moreover, it is doubtful that the quantity has ever exceeded its current values on the Earth's surface.

On the other hand, it is probable that a small, constantly replenished reservoir of condensed phosphates has existed throughout geological time. It should be recognized, however, that there is little tangible evidence to support the concept. At best, it can be claimed that the conditions should have allowed the formation of condensed phosphates by any one of several paths, and several known minerals are probably derived from condensed phosphates.

II.   THE FORMATION OF CONDENSED PHOSPHATES

One and only one single factor governs whether or not a phosphate is a condensed phosphate or an orthophosphate. If the M_2O - P_2O_5 ratio of a phosphate is less than 3, at least a part of the phosphate is present as a condensed phosphate (M is defined as one equivalent of either metal, hydrogen, or organic radical and is a unit of constitution of the phosphate.) (Van Wazer and Griffith, 1955). It can be demonstrated that the number of P - O - P linkages in a phosphate is also solely dependent upon the M_2O - P_2O_5 ratio of the phosphate, while the particular compound represented by the phosphate will depend upon the choice of M and the phase state of the composition.

Condensed inorganic phosphates may be prepared by one of three major routes. Phosphorus can be oxidized to phosphoric oxide and the phosphoric oxide may react with M_2O to yield a condensed phosphate. Phosphoric oxide is an ultraphosphate and is in the maximum possible condensation state of any phosphate.

A second means of forming a condensed phosphate is to remove M_2O from an orthophosphate to lower the M_2O - P_2O_5 ratio. This may be accomplished either by removing water from an acidic phosphate or metal oxide from a salt. It is also possible to effectively lower the M_2O - P_2O_5 ratio by a partial reduction of the metal, M, from a higher to a lower oxidation state.

A third means of preparing a condensed phosphate is by the substitution of oxygen for X in a compound containing P - X - P linkages, where X may be - S -, - Se -, - N -, or similar groups.

III. EXAMPLES OF THE FORMATION OF CONDENSED PHOSPHATES

The possible routes to specific condensed phosphates in nature are numerous. Perhaps the most probable of all condensed phosphates likely to have been encountered in the Precambrian environment was phosphoric oxide. Additionally, it is possible to form spontaneously most of the known inorganic phosphates from this compound. Phosphoric oxide is in the highest energy state of any phosphate, and, although formed readily, its lifetime as the P_4O_{10} molecule is expected to be very short in the presence of water or metal oxide. But the tetrametaphosphate, which forms spontaneously by the reaction of metal oxides, metal carbonates, or water with the P_4O_{10} molecule, is one of the more slowly reactive condensed phosphates toward hydrolytic degradation in dilute aqueous media near neutral pH values. It is more probable that the primitive seas contained tetrametaphosphate than trimetaphosphate, though the latter was probably present in munute quantities.

IV.   THE PREPARATION OF P_4O_{10}

As a result of the ease of formation of P_4O_{10} when elemental phosphorus is allowed to burn, laboratory work directed toward the formation of the compound by other routes has been relatively limited. It is known that SO_3 is capable of dislodging P_4O_{10} from calcium, magnesium, or alkali metal phosphates at temperatures no greater than $600^{\circ}C$. (Baumgarten and Brordenburg, 1939). It is also known that SiO_2 is capable of displacing P_4O_{10} from calcium phosphates at temperatures near 1000° (Aziev *et al.*, 1972). Little imagination is required to envision either volcanic action or the impact of lightning on igneous rocks generating temperatures far in excess of that required to volatilize P_4O_{10}. In fact, it is unlikely that lightning often strikes soil or rock without volatilizing some P_4O_{10}. If a reducing agent is present, even elemental phosphorus can be formed. The reaction paths of elemental phosphorus are more variable than P_4O_{10}., and a wide variety of products might be formed. Ultimately, both routes lead to orthophosphates.

V.   THE REMOVAL OF M_2O From Orthophosphate

Before considering the removal of the water of constitution from acidic phosphates other reactions should be discussed. Several reactions are known which are capable of producing P - O - P linkages from orthophosphates by the removal of metal oxides from the phosphate. The following reactions have been recognized for many years (Jamieson, 1842, von Knorre, 1900).

$$2Na_3PO_4 + 2NH_4Cl \rightarrow Na_4P_2O_7 + 2NaCl + H_2O + 2NH_3$$
$$\text{and}$$
$$3Na_4P_2O_7 + 6NH_4Cl \rightarrow 6NaCl + 2Na_3P_3O_9 + 3H_2O + 6NH_3.$$

More recently it has been shown that urea may also function in a similar manner (Griffith, 1975):

$$2MK_2HPO_4 + M(NH_2)_2CO \rightarrow 2(KPO_3)_M + MK_2CO_3 + 2MNH_3,$$
where the urea is not only removing water of constitution but also K_2O from the phosphate. The above reactions are probably all applicable to calcium phosphates, also yielding either $CaCl_2$ or $CaCO_3$.

One of the more interesting means of forming P - O - P linkages occurs when the metal of the metal oxide is reduced in oxidation state. If ferric phosphate is reduced to ferrous phosphate under anhydrous conditions, the resulting ferrous phosphate is a pyrophosphate (Hutter, 1953), while copper yeilds even metaphosphates.

$$2MCu_3(PO_4)_2 + 3MH_2 \rightarrow 2[CuPO_3]M + MCu_4P_2O_7 + 3MH_2O$$

$$2FePO_4 + H_2 \rightarrow Fe_2P_2O_7 + H_2O \ (460^\circ C).$$

Relatively low temperatures are required (about $400^\circ C$), and almost any reducing agent capable of reducing iron may be employed. It is even possible that P_4O_{10} could be prepared by reducing the iron to metallic iron in a reaction of this type if the iron is prevented from forming a carbonyl (Coulter, 1954):

$$4FePO_4 + 6CO \rightarrow 4Fe^\circ + P_4O_{10} + 6CO_2.$$

In the early preparation of elemental phosphorus in retorts, the phosphates were usually converted to lead salts before they were reduced. Lead phosphates have relatively low melting temperatures which aid in the reduction of the phosphate. Lead is easily reduced to the metal and this action releases P_4O_{10} for reduction to the elemental phosphorus. Charcoal was usually employed as a reducing agent.

VI THE REMOVAL OF WATER OF CONSTITUTION

Several means have been devised to prepare condensed phosphates from acidic orthophosphates. The most commonly employed method is to heat the phosphate to a temperature sufficiently high to drive water from the system. Numerous investigations have shown that urea and its hydrolysis products, cyanates, etc., function to remove water from an orthophosphate to form condensed phosphates. Sulfamide, $(NH_2)_2SO_2$, behaves in a similar manner to form ammonium sulfamate, $NH_4NH_2SO_3$, which decomposes near $160^\circ C$. Carbodimide is a powerful coupling agent for the formation of P - O - P linkages, and the compound is capable of dehydrating phosphates even to ultraphosphate structures (Glonek, 1974). Acetic anhydride has been employed in the low-temperature synthesis of condensed phosphates (Grunze, 1960). The author has shown that it is possible to form condensed phosphates from orthophosphoric acid merely by electrolyzing hydrogen and oxygen from the acid by applying a low potential dc current at room temperature (Griffith, 1958).

$$3H_3PO_4 \rightarrow H_5P_3O_{10} + 2H_2 + O_2$$

Another means of removing metals from phosphates occurs when CO is passed over a heated metal phosphate, if the metal is capable of forming a carbonyl. A reaction of this type has been observed, but the details have not been explored (Lockheart, personal communication):

$$Fe_3(PO_4)_2 + 6CO \rightarrow Fe_2P_2O_7 + Fe(CO)_5 + CO_2$$

Before discussing the complex minerals derived from pegmatites, the claim of Shepard will be considered (Shepard, 1878).
He claimed the existence of an anhydrous calcium pyrophosphate
which he believed to be of igneous origin. The primary evidence
for believing the phosphate to be of igneous origin is the fact
that it was anhydrous. If the sample analyzed by Shepard was
genuine, there can be little doubt that it was condensed phosphate. It is unfortunate that he could not divulge the precise
location of the deposit as a result of "economic" restraints.
It is only stated that the deposit was somewhere in the West
Indies. Even if the deposit ultimately proved to be of biological origin, the location of this deposit should be reestablished
for its scientific value, Shepard's claim is unique.

Two relatively recent works claim lomonosorite to contain
sodium metaphosphate (Gerosimovskii and Kozakova, 1962; Semenev
et al., 1961). The composition of the mineral is
$Na_3H_3MnTiSi_4O_{18}.2NaPO_3$ (metalomonosorite) or $Na_2Ti_2Si_2O_9.NaPO_3$
in lomonosorite. The evidence for the metaphosphate is not
absolute, but there is no doubt that on heating, a metaphosphate
could form even if the original salt were an orthophosphate.

Similar minerals are found in the deposits of Branchville,
Connecticut (Palache *et al.*, 1951). The minerals are reported
to have the following composition: $Mn_5H_2(PO_4)_4.4H_2O$ and
$(MnFeMg)H(PO_4).2H_2O$. If heated, condensed phosphates would
surely form. It is reasonable to assume that the source initially contained condensed phosphates and that they degraded during the period before solidification.

Deposits of both Na_3PO_4 and NaH_2PO_4 are known in the residues of salt lakes (McKelvey, 1972). The acidic phosphate is
of particular interest because its most probable source is a
condensed phosphate. The fact that it is not in the form of
microcosmic salt, $NaNH_4HPO_4$, does not exclude the possibility
it was derived from biological origin, but does cast doubt on
a biological source.

It is perhaps significant that none of the known phosphorus
minerals contains a sulfide group. Soluble sulfides are particularly effective in solubilizing phosphates of metals precipitated by sulfides. Iron, lead, copper, cobalt, nickel, and mercury phosphates can be quickly converted to a soluble phosphate
and an insoluble sulfide with $(NH_4)_2S$, Na_2S, or H_2S at room
temperature. Many of the standard laboratory preparations of
condensed phosphates rely on the sulfides. When a condensed
phosphate is solubilized it is possible for it to reorganize.
All polyphosphates with the exception of pyro- and triphosphate
are capable of spontaneously yielding trimetaphosphate by reor-
ganization in solution. In this instance, trimetaphosphate is
a more highly condensed phosphate than the polyphosphate from
which it is derived. The references to the phosphorylation

VII.   OXYGEN SUBSTITUTION REACTION

Two reactions of the oxygen substitution variety are worthy of consideration in the context of this analysis. It has been shown that P - NH - P linkages can be converted to P - O - P linkages merely by allowing water to react with the bond (Fluck, 1967). In the case of the cyclic imidophosphates, compounds of the NH group can be replaced by oxygen to yield trimetaphosphate:

Phosphorus pentasulfide, P_4S_{10}, can also react with water to yield condensed phosphates or thiophosphates and H_2S.

$$P_2S_5 + 7H_2O \rightarrow H_4P_2O_7 + 5H_2S$$

VIII.   THE DECOMPOSITION OF SALTS OF LOWER OXIDES

Condensed phosphates can be prepared by the disproportionation of the salts of the lower oxyacids of phosphorus. For example:

$$14MNaH_2PO_2 \rightarrow 6MPH_3 + 4MNa_2HPO_3 + 2[NaPO_3]M + MNa_4P_2O_7 + 3MH_2O$$

It is improbable that the above reaction has ever contributed a significant quantity of condensed phosphates to the environment because it is highly improbable that any significant quantity of reduced phosphorus has existed on Earth during geological time. There are no known mineral deposits of either phosphides or low oxides. The mineral deposits of the Earth have surely existed only as phosphates and almost exclusively as orthophosphates with but a minute fraction present as condensed phosphates.

IX.   CONDENSED PHOSPHATE MINERALS

There have been a few claims of condensed phosphate minerals. Whether or not these claims are as presented, they are worthy of consideration because they are, for the most part, claimed for minerals that are capable of having been derived from condensed phosphates, even if they are not presently of this structure. The more recent claims resulted from analyses of minerals derived from pegmatites.

reactions of the metaphosphates are well known.

X. CONCLUSION

There is no single large source of condensed phosphates on the surface of the Earth and there is small chance that a large source has ever existed. There are a wide variety of potentially useful, small abiotic sources which have probably replenished the Earth's store of condensed phosphates about as rapidly as they have degraded. The primary sources of condensed phosphates are probably biological; but since there is no known way to determine the origins of inorganic phosphates, the small quantities found in the environment may be partially derived from abiotic sources.

REFERENCES

Aziev, R. G., Vol'fkovich, S. N., Dubkina, G. A., and Mikhaleva, T. K. 1972. *Russian J. Phys. Chem.*, *46*:1.

Baumgarten, P., and Brandenburg. 1939.*Ber.*, *72B*:555.

Coulter, R. S. 1954. *Iron Age*, *173*:107.

Fluck, 1967. In *Topics in Phosphorus Chemistry*, John Wiley and Sons, New York, pg. 366.

Gerosimovskii, V. I., and Kazakova, M. E. 1962. *Doklady*, *142*:670.

Glonek, T., Van Wazer, J. R., Kulps, R. A., and Myers, F. C. 1974. *Inorg. Chem.*, *13*:2337.

Griffith, E. J. June 17, 1958, U. S. Patent No. 2,839,408.

Griffith, E. J. 1975. *J. Pure App. Chem.*, in press.

Grunze, I. E., Thilo, E., and Grunze, H. 1960. *Chem. Ber.*, *93*.

Hutter, J. C. 1953, *Ann. Chim.*, *8*:450.

Jamieson, A. 1842. *Ann. Pharm.*, *59*:350.

von Knorre, G. 1900. *Z. Anorg. Chem.*, *24*:381.

Lockheart, W. Personal communication.

McKelvey, V. E. 1972, *Environmental Phosphorus Handbook*, John Wiley and Sons, New York pg. 15.

Palache, C., Berman. H., and Frondell, C. 1951, *Dana's System of Mineralogy*, 7th ed., Vol. II, John Wiley and Sons, New York.

Semenev, E. I., Organova, N. I., and Kukharchik, M. W. 1961. *Kristallografiga*, *6*:925.

Shepard, U. 1878, *Amer. J. Sci. Arts*, *15*:49.

Van Wazer, R., Jr., and Griffith, E. J. 1955, *J. Amer. Chem. Soc.*, *77*:6140.

8

HYDROCARBONS AND FATTY ACIDS IN OIL SHALE OF PERMIAN IRATI FORMATION, BRAZIL

D. W. NOONER and J. ORO
University of Houston

The isoprenoid, normal, and polycyclic alkanes, the aromatic hydrocarbons, and the normal fatty acids from samples of two distinct (upper and lower) beds of oil shale in the Permian Irati Formation near Sao Mateus do Sul, Parana, Brazil, have been analyzed by gas chromatography and gas chromatography - mass spectrometry. Both the upper and lower beds yielded the same compounds (the amounts, given in ppm, are first, upper bed, and second, lower bed), i.e., isoprenoid alkanes (C_{16}, C_{18} - C_{21}; max = C_{19}; 196 and 152 ppm, respectively); normal alkanes (C_{12} - C_{29}; max = C_{17}, C_{12}; carbon preference index or CPI = 1.05; 95 and 73 ppm, respectively); polycyclic alkanes (C_{27} - C_{31} steranes and triterpanes; 188 and 100 ppm, respectively); aromatic hydrocarbons (member in the C_{12} - C_{23} range; 59 and 73 ppm, respectively); and normal fatty acids (C_{12} - C_{18}; max = C_{16}; CPI = 3.1; 30 and 48 ppm, respectively). The distribution of the alkanes extracted from the Irati Shale may be indicative of its depositional environment and of the biological material from which the organic matter in it was derived. The predominance of isoprenoid alkanes is an indication of high photosynthetic activity. Evidence of an algal contribution to this activity lies in the prevalence of n-heptadecane among the normal alkanes. Other possible contributors to the isoprenoids include ferns, crustacea, and bacteria. The relatively large amounts of triterpanes in the shale also suggest that algae, bacteria, and possibly ferns were significant sources of organic matter at time of deposition. Certain contemporary blue-green alga, bacteria, and ferns contain structures which could be modified by diagenesis and maturation to the triterpanes found in the oil shale.

I.   INTRODUCTION

An oil shale of high organic content occurs in the Permian Irati Formation of Brazil. The outcrop of the Irati Formation begins in Sao Paulo State and extends almost continuously in the form of a great "S" for 1700 km to the southern frontier of Brazil and into Uruguay (Padula, 1969). In parts of this range (southern Parana and Rio Grande do Sul) there are two distinct beds of oil shale which are separated by nonbituminous shale and limestone.

According to Beurlen (1953, 1955, and as reported by Padula, 1969) the organic matter in the shale was probably derived from fauna and flora proliferating in an intracontinental marine basin of reduced salinity. Fine-grained clastic material brought in by rivers and residues from the organisms formed the carbonate rocks and shale of the formation. The shale averages 20 - 30% organic matter (mostly kerogen), 60 - 70% mica and clay minerals, and 1 - 8% carbonate and silicate minerals (Padula, 1969). No average water content was given but one sample (from lower bed near Sao Gabriel) was reported to contain not more than 3% water.

We previously analyzed samples of the two beds (upper and lower) of oil shale and identified suites of isoprenoid and normal alkanes (Gibert *et al.*, 1975). For completeness, some of the data are included in this report. We also present the results of our study of the polycyclic alkanes, aromatic hydrocarbons, and fatty acids in organic matter extracted from the oil shale.

II.   EXPERIMENTAL

The samples of Irati Oil Shale were obtained from a drill hole located near Sao Mateus do Sul, Parana. They are from two beds of oil shale which are separated by a layer (8.6 m, average thickness) of nonbituminous shale and limestone. The upper bed has an average thickness of 6.5 m and averages 6.5% of pyrolytic oil; the lower bed averages only 3.2 m in thickness but yields an average of 9.1% pyrolytic oil. These were the same samples analyzed in our previous study (Gibert *et al.*, 1975).

The experimental methods are the same as given in our earlier report (Gibert *et al.*, 1975). An additional technique used (described below) is that of forming methyl ester derivatives of fatty acids. The eluates analyzed were the alkane, aromatic, and polar fractions, respectively, from the silica-gel chromatography of the bitumens extrated from the shale.

The isoprenoid and normal alkanes from the Irati Shale

were analyzed on a stainless steel capillary (152.4 m x 0.051 cm i.d.) coated with Polysev [*m*-bis-*m*-(phenoxyphenoxy)-benzene, Applied Science Laboratories].

The high molecular weight polycyclic alkanes in the branched-cyclic alkane fraction were analyzed on a stainless steel column (1.5 m x 0.2 cm i.d. packed with 3% SE-30 on diatomaceous earth). The branched-cyclic alkane fraction was obtained by removing normal alkanes from the *n*-hexane eluate as urea complexes (Zimmerschied *et al*., 1949). The aromatic hydrocarbons in the benzene eluate were analyzed using the Polysev capillary and SE-30 packed columns.

The fatty acids from the oil shale were analyzed as methyl ester derivatives (Nooner *et al*., 1973). The methanol eluates were saponified with alcoholic potassium hydroxide and then extracted with hexane to remove impurities. The alkaline solution was then neutralized with hydrochloric acid and the fatty acids were extracted with hexane. After removal of water the acids were heated (reflux) in a mixture of hexane and methanol to which 1 ml of boron trifluoride-methanol reagent (Applied Science Laboratories) had been added. After reaction, water was added and the methyl esters of the fatty acids were extracted with hexane. The hexane extracts were concentrated and the fatty acid esters contained therein were analyzed on the capillary (Polysev) column.

All of the aforementioned analyses were made on a Varian 1200 FID gas chromatograph and an LKB 9000 gas chromatograph-mass spectrometer combination (modified by replacing the original oven with a Perkin-Elmer 990 gas chromatograph).

III. RESULTS

The isoprenoid, normal, and polycyclic alkanes, aromatic hydrocarbons, and fatty acids of the upper and lower beds of the Irati Oil Shale are qualitatively the same. Therefore, gas chromatographic traces of material extracted from either bed are used to illustrate the results obtained, e.g., Figs. 1a and 1b, alkanes from lower bed; Fig. 1c, fatty acids from lower bed; Fig. 2, aromatic hydrocarbons from upper bed. The semiquantitative data for both beds of oil shale are summarized in Table 1.

The mass spectra obtained as part of this study have not been reproduced in this report. References to similar published spectra are indicated in the following sections.

A. *Isoprenoid and Normal Alkanes*

The distribution profile of isoprenoid and normal alkanes from the Irati Shale is shown in Fig. 1a. The normal alkanes above C_{23} are not shown; see Gibert *et al*.(1975) and Table 1

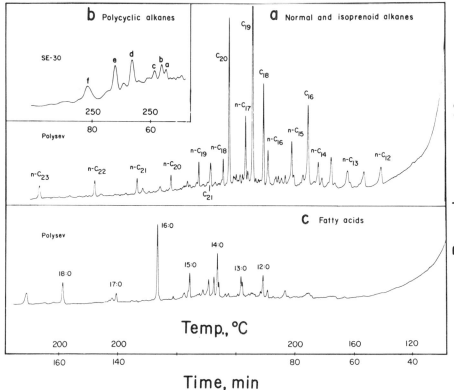

Fig. 1. *Alkanes and fatty acids from the lower bed, Irati Oil Shale (Parana, Brazil). (a) Normal and isoprenoid alkanes. Instrument: Varian 1200. Attenuation 1, range 10. GLC column: 152.3 m x 0.051 cm i.d. stainless steel, 5% Polysev (7-ring m-polyphenyl ether), 703 g/cm^2 of helium. Temperatures: Isothermal at 60°C for 10 min, then programmed to 200°C at approximately 2° per minute. About 1/50 injected. (b) Polycyclic alkanes (polycyclic portion of the branched-cyclic alkanes chromatogram). Instrument: Varian 1200. Attenuation 2, range 10. GLC column: 1.5 m x 0.2 cm i.d. stainless steel packed with 3% SE-30 on Varaport 30, 1405 g/cm^2 of helium. Temperatures: Programmed from 60 - 250°C at approximately 4° per minute. About 1/20 injected. (c) Fatty acids (as methyl esters). Instrument, GLC column and temperatures: Same as in (a). About 1/10 injected.*

for the complete distribution of normal paraffins. Except for maxima at C_{12} and C_{17}, the normal alkanes have a smooth distribution. The chromatogram (Fig. 1a) shows that the peaks corresponding to pristane (C_{19}), phytane (C_{20}), and norpristane (C_{18}) are, in this order, the largest. On the other hand, while the

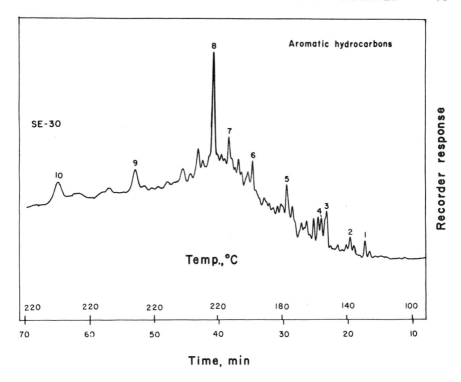

Fig. 2. Aromatic hydrocarbons from the upper bed, Irati Oil Shale (Parana, Brazil). Instrument: Varian 1200. Attenuation 1, range 100. GLC column: 1.5 m x 0.2 cm i.d. stainless steel packed with 3% SE-30 on Varaport 30, 1405 g/cm² of helium. Temperatures: Programmed from 60 - 220°C at approximately 4° per minute. About 1/2 injected.

C_{16} isoprenoid is relatively abundant, the C_{17} isoprenoid is either absent or extremely small. The absence of C_{28} pentacyclic triterpanes in sedimentary samples is a similar situation (Kimble et al., 1974).

B. Polycyclic Alkanes

The section of the gas chromatogram of the branched-cyclic fraction showing the polycyclic alkanes is presented as Fig. 1b. Peaks a, b, and c represent mixtures of compounds. The mass spectrum taken as peak a was eluted had molecular ions at m/e 372, 386, and 400. The predominance of the ion at m/e 217 along with a rather large ion at m/e 191 indicated that the compounds were probably steranes and tetracyclic triterpanes (Gallegos, 1971). Assignment of structural type is shown in Table 2. The

Table 1
Isoprenoid, Normal, and Polycyclic Alkane, Aromatic Hydrocarbon, and Fatty Acid Content of the Irati Oil Shale

Oil shale bed	Hydrocarbons				Normal fatty acids[a]	
	Alkanes			Aromatics		
	Isoprenoid[b] (ppm)	Normal[c] ppm CPI[d]		Polycyclic (ppm)	(ppm)	ppm CPI[d]
Upper	196	95	1.05	188	59	30 3.1
Lower	152	73	1.05	100	73	48 3.1

[a] Range: C_{12} -C_{18}; max: C_{16}.

[b] Range: C_{16} - C_{21}; max: C_{19}.

[c] Range: C_{12} - C_{29}; max: C_{12}, C_{17}.

[d] Carbon preference index (Bray and Evans, 1961; Kvenvolden, 1966); in the C_{18} - C_{29} range for *n*-alkanes.

mass spectrum of the mixture represented by peak b also had molecular ions at m/e 372, 386, and 400 and was very similar to the spectrum for peak a. See Table 2 for relative intensities of the molecular ions.

Peak c was shown to result from a four-component mixture as evidenced by molecular ions at m/e 370, 386, 400, and 414. The high intensity of the ion at m/e 217 and the molecular ion at m/e 370 indicates that the major component in peak c is a pentacyclic triterpane. When the ions at m/e 386, 400, and 414 are deleted, the spectrum obtained for peak c is almost identical to the spectrum of "Triterpane A" (Douglas *et al.*, 1969), which has a molecular ion at m/e 370 and a major fragment ion at m/e 191. The other compounds (molecular ions at m/e 386, 400, and 414) are probably steranes and tetracyclic triterpanes. Definite structural types cannot be assigned.

Peaks d and e represent essentially single compounds rather than mixtures; peak f represents a two-component mixture. The ion at m/e 191 is very large in all cases (the most intense fragment ion for the compounds represented by peaks e and f), while the ion at m/e 217 is small. This observation together with molecular ions of m/e 398, 412, and 426, respectively, indicate that the compounds of peaks d and e and the two components of peak f are pentacyclic triterpanes. The mass spectrum obtained as peak d was eluted is similar to the spectra of "Triterpane B" (Douglas *et al.*, 1969) and the compound of "Peak 30" (Gallegos, 1971). The mass spectrum of the compound eluted as peak e is virtually identical to the spectrum of "Triterpane C," as reported by Douglas *et al.* (1969). Both of these spectra match in important detail the spectra of "Compound 48" (Anders

Table 2
Polycyclic Alkanes in the Irati Oil Shale

Peak (Fig. 1b)	Empirical formula	Molecular ion	Relative intensities (%) of molecular ion	Structural type
a	$C_{27}H_{48}$	372	27.6[a]	Sterane
a	$C_{28}H_{50}$	386	13.8[a]	--
a	$C_{29}H_{52}$	400	34.5[a]	Tetracyclic triterpane
b	$C_{27}H_{48}$	372	35.6[a]	Sterane
b	$C_{28}H_{50}$	386	8.9[a]	--
b	$C_{29}H_{52}$	400	20.0[a]	Tetracyclic triterpane
c	$C_{27}H_{46}$	370	12.0[b]	Pentacyclic triterpane
c	$C_{28}H_{50}$	386	9.4[b]	--
c	$C_{29}H_{52}$	400	8.6[b]	--
c	$C_{30}H_{54}$	414	8.4[b]	--
d	$C_{29}H_{50}$	398	20.8[b]	Pentacyclic triterpane
e	$C_{30}H_{52}$	412	16.4[b]	Pentacyclic triterpane
f	$C_{30}H_{52}$	412	6.1[b]	Pentacyclic triterpane
f	$C_{31}H_{54}$	426	9.8[b]	Pentacyclic triterpane

[a]Relative to intensity of ion at *m/e* 271.
[b]Relative to intensity of ion at *m/e* 191.

and Robinson, 1971). and the compound of "Peak 31" (Gallegos, 1971).

The compound eluted as the major component of peak f (*m/e* 426) was not reported in the Westwood Shale (Douglas *et al.*, 1969), an oil shale of Carboniferous age (250 - 300 x 10^6 years old). This triterpane, based on comparison of mass spectra, appears to be similar in structure to the triterpane of "Peak 33" (Gallegos, 1971).

C. Aromatic Hydrocarbons

The aromatic hydrocarbon profile for the Irati Oil Shale

is shown in Fig. 2. The numbered peaks are those for which
essentially single-component mass spectra were obtained. These
spectra are similar to some of those published by Anders *et al.*
(1973), in their study of the aromatic hydrocarbons in the Green
River Oil Shale. The numbered peaks of Fig. 2 are identified
in Table 3. The empirical formulas were deduced from the mass
spectra and retention times of the particular compounds.

The missing series, e.g., C_nH_{2n-6} and C_nH_{2n-10}, and the
missing members of series present are believed due to the small
sample size (0.5 g) used and not to their absence. However,
one series detected in the Irati Oil Shale (C_nH_{2n-20}) may not
be present in the Green River Oil Shale. This more highly aromaticized material may be an indication that the Irati Oil Shale
is more highly matured than the Green River Oil Shale, probably
because of its significantly greater age.

The arguments advanced by Anders *et al.* (1973) as to the
possible origin of the aromatic hydrocarbons in the Green River
Oil Shale hold for the origin of the aromatic hydrocarbons in
the Irati Oil Shale. The aromatic hydrocarbons in both cases
appear to be derived from naturally occurring isoprenoid lipids.
The higher molecular weight aromatic hydrocarbons may be aromaticized polycyclic alkanes, the series depending upon the degree
of aromatization and the particular hydrocarbons aromaticized.

D. Fatty Acids

The gas chromatography trace of normal fatty acids (as
methyl esters) from the Irati Oil Shale is shown in Fig. 1c.
These fatty acids are in the C_{12} - C_{18} range with maximum at
C_{16}. The peak adjacent to peak 14:0 is caused by the methyl
ester of C_{16} isoprenoic fatty acid. See Van Hoeven (1969) for
mass spectra of the type obtained in this study. The other non-
designated peaks have not been identified. The shoulders on
peaks 13:0 and 14:0 represent compounds that make little contribution to the mass spectra obtained on the LKB 9000.

The fatty acids of Fig. 1c were derived from the oil shale
by extraction. As with the Green River Oil Shale (Burlingame
et al., 1969; Douglas *et al.*, 1968), a larger sample size, de-
mineralization prior to extraction, and/or oxidation would probably result in the recovery of a wider range of normal fatty
acids along with additional classes of acids.

IV DISCUSSION

The Irati Formation belongs to the Estrada Nova Group which
contains fauna and flora resembling that of the Lower Beaufort
Series of the Karoo System of South Africa (Padula,1969). The
fauna include fish, crustacea, and reptilia; the flora Pteridophyta and Thallophyta. The extracted isoprenoid and normal

Table 3
Aromatic Hydrocarbons in the Irati Oil Shale

Peak (Fig. 2)	Empirical formula	Molecular ion	Series
1	$C_{13}H_{18}$	174	C_nH_{2n-8}
2	$C_{12}H_{12}$	156	C_nH_{2n-12}
3	$C_{14}H_{20}$	188	C_nH_{2n-8}
4	$C_{13}H_{14}$	170	C_nH_{2n-12}
5	$C_{14}H_{16}$	184	C_nH_{2n-12}
6	$C_{17}H_{22}$	226	C_nH_{2n-12}
7	$C_{18}H_{24}$	240	C_nH_{2n-12}
8	$C_{18}H_{22}$	238	C_nH_{2n-14}
9	$C_{23}H_{34}$	310	C_nH_{2n-12}
10	$C_{21}H_{22}$	274	C_nH_{2n-20}

alkanes and steriod and triterpenoid alkanes may provide clues as to the origin of the organic matter in the Irati Shale, i.e., which of the aforementioned fauna and flora are significant precursors.

The predominance of the n-C_{17} alkane among the normal alkanes indicates an algal contribution (Han *et al.*, 1968; Eglinton, 1969; Han and Calvin, 1969; Gelpi *et al.*, 1970). The large amounts of isoprenoid alkanes may indicate high photosynthetic activity since these hydrocarbons can be produced by the diagenesis of the phytyl moiety of chlorophyll. However, the presence of isoprenoid alkanes is not necessarily exclusively linked to chlorophyll since pristane and phytane have been found in several anaerobic nonphotosynthetic bacteria (Han and Calvin, 1969). According to Albrecht and Ourisson (1969), the predominance of pristane in a suite of isoprenoid alkanes from an ancient sediment indicates a marine origin. This would support the proposal of Beurlen (1953, 1955) regarding the depositional environment of the oil shale. However, definite conclusions are not possible because Mathews *et al.*(1972) report in their study of a fresh water section of the Evergreen Shale of Australia that pristane was dominant.

A striking feature of the polycyclic alkanes in the Irati Shale is that triterpanes are present in at least a 4:1 excess over steranes. This indicates a predominantly phytological rather than a possible zoological origin of the organic matter in the Irati Shale. The polycyclic alkane analyses cannot definitely establish whether megaflora or microflora was the greater contributor to the organic matter, since triterpane structures have been found in contemporary specimens of both; e.g., in Pteridophyta (Berti and Bottari, 1968; Bottari et al., 1972 and in Thallophyta (De Rosa et al., 1971, 1973; Gelpi et al., 1970). However, if deposition did occur in a shallow intracontinental marine basin (Beurlen, 1953, 1955), algae are probably the main precursors of the organic matter in the Irati Oil Shale.

Acknowledgment

We thank G. Holzer and W. Horine for assistance in obtaining mass spectra and J. M. Gibert and I.M R. de Andrade Bruning for technical assistance. This work was supported in part by grants from the National Aeronautics and Space Administration. One of us (J. Oro) wishes to thank the National Aeronautics and Space Administration for a 1-year appointment as NASA Life Scientist in Chemistry at Ames Research Center, Moffett Field, California Interchange Agreement NCA 2-OP295-501.

REFERENCES

Albrecht, P., and Ourisson, G. 1969. *Geochim. Cosmochim. Acta,* *33*:138.

Anders, D.E., Doolittle, F. G., and Robinson, W. E. 1973. *Geochim. Cosmochim. Acta,* *37*:1213.

Anders, D. E., and Robinson, W. E. 1971. *Geochim. Cosmochim. Acta, 35*:661.

Berti, C., and Bottari, F. 1968. *Progr. Phytochem., 1*:589.

Beurlen, K. 1953. *Divisao de Geol. e Miner. Rio de Janeiro, Avulso 59.*

Beurlen, K. 1955. *Brazil Div. Geol. e Miner. Bol., no. 153.*

Bottari, F., Marsili, A., Morelli, I., and Pacchiani, M. 1972. *Phytochemistry, 11*:2519.

Burlingame, A. L., Haug, P. A., Schoes, H. J., and Simoneit, B. R. 1969. In P. A. Schenck and I. Havenaar (eds.), *Advances in Organic Geochemistry, p. 85.* Pergamon Press, New York.

De Rosa, M., Gambacorta, A., Minale, L., and Bul'ock, J. D. 1971. *Chem. Commun.,* p. 619.

De Rosa, M., Gambacorta, A., Minale, L., and Bul'ock, J. D. 1973. *Phytochemistry, 12*:1117.

Douglas, A. G., Douraghi-Zadek, K., Eglinton, G., Maxwell, J. R., and Ramsay, J. N. 1968. In G. C. Hobson and G. D. Speers (eds.), *Advances in Organic Geochemistry.* p. 315. Pergamon Press, New York.

Douglas, A. G., Eglinton, G., and Maxwell, J. R. 1969. *Geochim. Cosmochim. Acta, 33*:579.

Eglinton, G. 1969. In G. Eglinton and M. T. J. Murphy (eds.), *Organic Geochemistry,* p. 20. Springer, New York.

Gallegos, E. J. 1971. *Anal. Chem., 4*:1151.

Gelpi, E., Schneider, H., Mann, J., and Oro, J. 1970. *Phytochemistry, 9*:603.

Gibert, J. M., de Andrade Bruning, I.M.R., Nooner, D. W., and Oro, J. 1975. *Chem. Geol., 15*:209.

Han, J., and Calvin, M. 1969. *Proc. Natl. Acad. Sci. U.S.A., 64*:436.

Han, J., McCarthy, E. D., and Calvin, M. 1968. *J. Chem. Soc. C,* p. 2784.

Kimble, B. J., Maxwell, J. R., Philp, R. P., Eglinton, G., Albrecht, P., Ensminger, A., Arpino, P., and Ourisson, G. 1974. *Geochim. Cosmochim. Acta, 38*:1165.

Kvenvolden, K. A. 1966. *Nature (London), 209*:573.

Mathews, R. T., Igual, X. P., Jackson, K. S., and Johns, R. B. 1972. *Geochim. Cosmochim. Acta, 36*:885.

Nooner, D. W., Updegrove, W. S., Flory, D. A., Oro, J., and Mueller, G. 1973. *Chem. Geol., 11*:189.

Padula, V. T. 1969. *Bull. Amer. Assoc. Petrol. Geol., 42*:591.

Van Hoeveven, W. 1969. Ph. D. Dissertation, University of California (Berkeley), Berkeley, California.

Zimmerschied, W. J., Dinerstein, R. A., Weitkamp, A. W., and Marschner, R. F. 1949. *J. Amer. Chem. Soc., 71*:2947.

9

THE STABLE ISOTOPES OF HYDROGEN IN PRECAMBRIAN ORGANIC MATTER

THOMAS C. HOERING
Carnegie Institution of Washington

The δD of ancient petroleums and extracts of Precambrian sediments range from -69.2 to -163.7. While these values overlap the range found in the lipid fraction of living plants, it is likely that processes have occurred in ancient sediments which exchange organically bonded hydrogens. Any record of Precambrian photosynthesis contained in the hydrogen isotopes of the original organic matter may be obscured.

I. INTRODUCTION

Studies on the distribution of the stable carbon isotopes have given valuable information on the history of the organic matter in Precambrian-aged rocks (McKirdy, 1974). The corresponding distribution of the hydrogen isotopes is not known. As part of our program on the biogeochemistry of deuterium, measurements have been made on ancient petroleums and rock extracts.

---

[1]The absolute abundance of ^{2}H (D) in sea water is 155 parts per million. Results here are reported in terms of parts per thousand difference in the D/H ratio as compared to Standard Mean Ocean Water distributed by the International Atomic Energy Agency, Vienna, Austria:

$$\delta D = \frac{(D/H)_x - (D/H)_s}{(D/H)_s} \times 1000$$

where the subscripts x and s refer to the unknown and the standard.

During photosynthesis, plants obtain their organically bonded hydrogen from the surrounding waters. There is a large isotope effect in the process (Smith and Epstein; Hoering, 1974, 1975)[1] Living plants are markedly depleted in the heavy isotope.[1] Thus, if the organically bonded hydrogen in the organic matter of sedimentary rocks is stable over long periods of geological time, measurements of hydrogen isotope ratios could give information on ancient photosynthetic processes and on the D/H ratio in waters of ancient oceans.

However, sedimentary rocks are permeated with water during their history and it is not known how extensively the water interacts with sedimentary organic matter and exchanges its bonded hydrogen.

The purpose of this chapter is to measure the hydrogen isotope ratio in selected samples of ancient organic matter and to try to decide between the two possibilities.

II. EXPERIMENTAL

Samples of crude petroleums from producing wells were obtained from Dr. Wilson Orr of the Mobil Field Research Laboratory, Dallas, Texas. The oil seep from the Nonesuch Shale was collected in the White Pine Copper Mine, White Pine, Michigan (Barghoorn, 1965). Portions of the oils were placed on a silica-gel, chromatography column and eluted successively with 1.5 vol of hexane, benzene, and methanol. The solvents were evaporated in a vacuum, rotating evaporator. The sedimentary rocks were collected by the author, either as outcrop samples or as diamond drill cores. The surfaces were scraped and thoroughly washed prior to crushing and ball milling. The rock powders were extracted with benzene - methanol mixtures (1/1 vol) using ultrasonic energy. After evaporating the solvent, the extracts were fractionated by silica gel chromatography. Previous studies (Hoering, 1967) indicated that the extractable organic matter in the rock specimens used here are probably in place and are not contamination.

Approximately 15 mg of organic matter was combusted quantitatively to carbon dioxide and water. The water was trapped and converted to molecular hydrogen by reaction with uranium metal at 700°C (Friedman and Hardcastle, 1970). Details of the combustion procedure and the isotopic analysis of hydrogen by mass spectrometry have been published (Hoering, 1974, 1975). Results of the analysis of sedimentary organic matter are shown in Table 1. For comparison, the results of hydrogen isotope measurements on fractions from living plants are given in Table 2 (Hoering, 1974, 1975).

III. DISCUSSION

TABLE 1
Hydrogen Isotope Content of Petroleums and Rock Extracts

Sample	δD total[a]	δD hexane eluate	δD benzene eluate	δD methanol eluate	Location	Age
798	-153.3	-165.0	-159.4	-159.0	Midway-Sunset Field, California	U. Pliocene
Green River Shale	-187.3	-234.6	-215.6	-150.0	Parachute Creek, DeBeque, Colorado	Eocene
801	-142.5	-14714	-132.7	-148.5	Midway-Sunset Field, California	L. Miocene
799	-150.5	-163.7	-151.5	-158.5	Midway-Sunset Field, California	U. Miocene
4159	-121.1	-128.0	-113.2	-121.2	Santa-Fe Springs, California	U. Miocene
3944	-105.9	-109.5	-89.5	-112.6	Lab-E-Safid, Iran	Miocene
4190	-109.4	-112.1	-112.5	-103.9	Deep River, Michigan	Devonian
Nonesuch Shale	-69.2	-68.8	-64.7	-65.0	White Pine Copper Mine, Michigan	1100 million years
McMinn Shale	-111.5	-115.8	-116.9	-100.7	N.Territory, Australia	1400 million years
Muhos Shale	-117.2	n.d.[b]	n.d.	n.d.	Finland	1300 million years
Barney Creek Shale	-97.1	-102.4	-94.1	-83.2	McArthur River N. Territory Australia	1600 million years

[a]Units of δD are defined in Footnote 1.
[b]n.d.: no data

The data in Table 1 show no obvious differences in δD with geological time, with the exception of the White Pine Petroleum seep. No explanation can be given at present for the exception. The range of δD of the petroleums and rock extracts overlaps the corresponding range of plant lipid material, as shown in Table 2. Lipids are the presumed source of much of the extractable organic matter and petroleums.

Further interpretation is limited by our lack of understanding of the chemical pathways of organic hydrogen during long periods of geological time. The possibility exists that the water in rocks can interact with organic matter and change the isotopic content of originally deposited material.

There are indications that such exchange occurs. The organic matter in the Green River Shale (Table 1) has had a very low temperature history and the rock contains an immature suite of saturated hydrocarbons. There is a wide variation in the δD of the classes of compounds separated by silica-gel chromatography. In contrast, older rocks, which have had a longer and higher temperature history, show a much more homogeneous distribution of hydrogen isotope contents. This result is to be expected if the old organic matter had the opportunity to exchange with a large volume of water.

Previous work in this laboratory (Hoering, 1968) showed that when saturated hydrocarbons were produced by mild thermal treatment of kerogen which had been exposed to pure D_2O, there was an appreciable amount of excess deuterium in the newly formed hydrocarbons. Extensive interaction with water is indicated. It is not known whether the effect is due to homogeneous exchange of hydrocarbons with water or if the kerogen abstracted hydrogen from the water during the chemical processes which formed the hydrocarbons.

Other experiments could give evidence on the stability of organically bonded hydrogen. Simulated diagenesis or sedimentary organic matter, using deuterium-enriched water, under hydrothermal conditions could measure the rates of isotopic homogenization and equilibrium distribution of the isotopes among various classes of organic compounds.

Table 2 shows that the lipids of plants fall into two groups based on their hydrogen isotope ratios. The first group, containing the fatty acids and hydrocarbons, is synthesized via the two-carbon acetate pathway and contains a larger D/H ratio than the second group which is synthesized via the five-carbon isoprenoid pathway. Normal and isoprenoid hydrocarbons were isolated from the hexane eluate of the extract from the Green River hale (Table 1) by urea and thiourea adductination. The δD of the two fractions were measured and found to be -164.2 for the normal hydrocarbons and -227.5 for the isoprenoid hydrocarbons The difference between the two is similar to that found for the lipid classes of plants. A measurement of the δD of co-

TABLE 2
Hydrogen Isotope Content of Fractions from Plants[a]

	PAJ-2[b]	PAJ-1[c]	LGL[d]	DEL-6[e]
Water	-7.5	-7.5	-20.0[f]	-6.0
Extracted plant	-112.7	-156.1	-101.9	-82.5
Total saponified liquid	-155.0	-110.4	-140.7	-134.7
Total nonsaponified liquid	-271.0	n.d.	-192.0	-178.9
Saturated fatty acids	-129.9	-114.5	-131.3	-127.7
Unsaturated fatty acids	-146.3	-126.1	-143.1	-118.7
Saturated hydrocarbons	-141.3	-131.7	-146.1	-158.1
Carotene pigments	-267.6	n.d.	n.d.	-177.3
Phytol	-308.9	-337.6	-207.2	-211.1
Sterols	n.d.	n.d.	-233.2	-209.6

[a]Units of δD are given in Footnote 1.

[b]PAJ-2: red alga, *Hypnea cornuta*, Gulf of Mexico, Port Aransas, Texas.

[c]PAJ-1: green alga, *Ulva fasciata*, Gulf of Mexico, Port Aransas, Texas.

[d]LGL: honey locust, *Gledititsia triacanthus*, Geophysical Laboratory grounds, Washington, D.C.

[e]DEL-6: marsh grass, *Spartina pataans*, The Great Marsh, Lewes, Delaware.

[f]Waters distilled from freshly harvested leaves, Ground waters have δD of -39.0.

existing normal and isoprenoid hydrocarbons in ancient rocks would indicate the extent of isotope homogenization.

Determination of hydrogen isotope ratios in Precambrian kerogens would be more meaningful, but it is difficult to obtain kerogen free of hydrous minerals or hydroscopic salts. Measurements on massive, mineral-free, carbonaceous deposits such as the anthraxolites or thucholites, which are widely distributed in the Precambrian, would be useful.

While not conclusive, current evidence indicates that the hydrogen isotope ratio in the extractable organic matter of Precambrian rocks is controlled by isotope-exchange reactions with water. Any record of the isotopes in the originally sedimented organic matter has been obscured.

REFERENCES

Barghoorn, E. S., Meinschein, W. G., and Schopf, J. W. 1965. *Science, 148:*461.

Friedman, I., and Hardcastle, K. 1970. *Geochim Cosmochim. Acta, 34:*125.

Hoering, T. C. 1967. *Researches in Geochemistry,* Vol. 2, pp. 89 - 111.

Hoering, T. C. 1968. *Carnegie Inst. Yearb. 67:*1199.

Hoering, T. C. 1974. *Carnegie Inst. Yearb. 73:*590.

Hoering, T. C. 1975. *Carnegie Inst. Yearb. 74:*598.

McKirdy, D. M. 1974. *Precambrian Res. 1:*75.

Smith, B. N., and Epstein, S. *Plant Physiol., 46:*738.

10

EVOLUTION OF THE TERRESTRIAL OXYGEN BUDGET

MANFRED SCHIDLOWSKI AND RUDOLF EICHMAN
Max-Planck-institut für Chemie

Based on a geochemical ^{13}C mass balance and current concepts for the increase of the stationary sedimentary mass as a function of time, a quantitative model is proposed for the evolution of photosynthetic oxygen. According to this model, a stationary reservoir of sedimentary organic carbon close to 80% of the present one should have already existed some 3.3×10^9 years ago , and thus, by inference, a stoichiometric oxygen equivalent of photosynthetic pedigree. As geologic evidence indicates an oxygen-deficient environment during the Early and most parts of the Middle Precambrian, we must necessarily assume that the partitioning of this oxygen between the "bound" and "molecular" reservoirs was different from that of today, with all photosynthetic oxygen produced being instantaneously sequestered by the crust as a result of very effective O_2-consuming reactions. Therefore, the reservoir of sedimentary organic carbon as derived from our model is just a measure of the gross amount of photosynthetic oxygen released, withholding any information as to how this oxygen was partitioned between the principal geochemical reservoirs. The virtual constancy of the sedimentary carbon isotope record would, furthermore, testify as to an extremely early origin of life on Earth, since the impact of organic carbon on the geochemical carbon cycle can be traced back almost 3.5×10^9 years.

I. INTRODUCTION

It is generally accepted by now that the buildup of the terrestrial free oxygen budget did, in all probability, not re-

sult from inorganic photolysis of water vapor in the upper atmosphere (with preferential escape of hydrogen into space), but
rather from an organic photochemical effect bringing about the
same result in a biologically catalyzed thermodynamic "uphill"
reaction (cf. Rabinowitch, 1945; Broecker, 1970; Welte, 1970;
Junge *et al.*, 1975; and others). Such an effect could have
started the operation when, during organic evolution, mutants
made their appearance, which disposed of the amino-heterotrophic
way of life entertained by their abiotic ancestors, having a-
quired instead the capability to directly synthesize organic
substances from inorganic compounds of their environment. A
convenient contrivance serving this purpose was the reduction
of carbon dioxide to carbohydrates which, from the standpoint
of energetics, is effected most economically by photosynthetic
activity of green plants and certain blue-green algae (Broda,
1975). From carbon dioxide and water these "autotrophic" life
forms were able to synthesize organic matter, with molecular
oxygen released as a metabolic by-product.

Quantitatively, the process of photosynthesis is the most
important biochemical reaction taking place on Earth. It is,
therefore, not unreasonable to assume that the oxygen content
of the terrestrial atmosphere is ultimately due to photosynthetic O_2 production. The oldest paleontological evidence of biogenic oxygen production is provided by algal bioherms from the
Bulawayan Group of Rhodesia (MacGregor, 1940; Schopf *et al.*,
1971.), previously assumed to have a minimum age of some 2.9 x
10^9 years (Bond *et al.*, 1973), which, however, has shifted late-
ly to about 2.6 - 2.7 x 10^9 years (Hawkesworth *et al.*, 1975).

II.   GEOCHEMICAL CARBON CYCLE AND TERRESTRIAL OXYGEN BUDGET

Recently, we have proposed a quantitative model for the
evolution of the terrestrial oxygen budget of photosynthetic
pedigree based, *inter alia,* on the isotopic geochemistry of
sedimentary carbon (Schidlowski *et al.*, 1975; Junge *et al.*,
1975). A necessary prerequisite for the following considerations is some basic information on the terrestrial carbon cycle as provided by the box model of Fig. 1.
Figure 1 shows the principal terrestrial carbon reservoirs
of atmosphere - ocean, biosphere, crust, and mantle along with
the transfer channels between the individual boxes. The primary source of all carbon now stored in reservoirs A, B, and C
is, naturally, the mantle reservoir D, the material having
been originally transferred via channel F_{DA} into box A. According to recent budget calculations (cf. Ronov, 1968; Li, 1972;
Broecker, 1973), box A is, however, very small, containing a-
bout one per mil only of all carbon which has ever crossed the

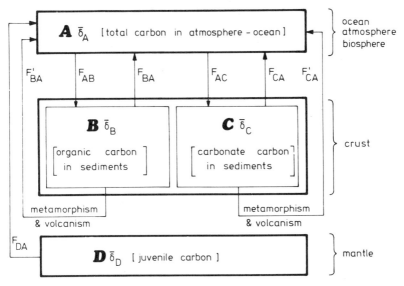

Fig. 1.   Box model of the terrestrial carbon cycle showing principal reservoirs and fluxes (adapted from Holland, 1976). $\bar{\delta}_A$ $\bar{\delta}_B$, etc., denote average $\delta^{13}C$ values of reservoirs.

crust - mantle boundary.   Since A is, in all probability, mainly stabilized by marine silicate equilibria (Sillén, 1961), it should have kept its size--at least approximately--from those remote times when the oldest oceans appeared on the surface of the Earth.   This would imply that the continuous carbon influx into A must have been offset by a corresponding efflux into the two crustal boxes B and C, storing carbon as either organic carbon or carbonate carbon.

As for reservoir D, there is reason to believe that, since 4.5×10^9 years have elapsed since the start of the degassing process, this primary mantle source is practically dried up now (with the exception perhaps of a certain fraction having remained in a virtually immobile, noninteractive state).   If we get any supplies from reservoir D now, it is likely to be recycled material ploughed down several hundred million years ago in some subduction zone.

Assuming that terrestrial free oxygen owes its origin to oxygen-releasing photosynthesis, its reservoirs and fluxes must be coupled with the carbon cycle as outlined in the box model of Fig. 1.   The crucial quantitative relationship which bears on the following model exists between the overall oxygen budget of photosynthetic derivation and box B, i.e., the totality of organic carbon buried within the sediments of the Earth's crust.

From a simplified version of the photosynthesis reaction,

$$CO_2 + H_2O \xrightarrow{h\nu} CH_2O + O_2,$$

it is apparent that for each carbon atom incorporated in organic matter one O_2 molecule will be released to the environment. As we know from our box model, the largest reservoir of organic carbon is within the crust (about 10^{22} g, Ronov, 1968), the living biomass ($\sim 10^{18}$ g; Garrels and Mackenzie, 1972) being smaller by four orders of magnitude. If we could trace, therefore, through geologic time the reservoir of sedimentary organic carbon--that is, the size of our reservoir B--then we would get, at the same time, the stoichiometric oxygen equivalent of photosynthetic origin. By introducing two adroit geochemical tricks, it is indeed possible to get at least a reasonable approximation for the evolution of reservoir B and, accordingly, of the terrestrial oxygen budget of photosynthetic pedigree.

III.   TERRESTRIAL ^{13}C MASS BALANCE

The first trick is provided by the overall terrestrial ^{13}C mass balance. From the isotopic composition of deep-seated carbon (diamonds, carbonatites, and carbonate constituents of kimberlites), we may infer that primordial mantle carbon has an average δ^{13}C value close to $-5^o/oo$ vs PDB (Deines and Gold, 1973). Since there is no reason to assume that fractionation processes have occurred during the transfer of this mantle carbon to the crust, the average δ^{13}C value of crustal carbon must be $-5^o/oo$ too. Now we are aware that crustal carbon comprises two carbon species, namely C_{org} and C_{carb}. As a result of a kinetic fractionation effect which governs biological carbon fixation, the crustal carbon box as a whole with a δ^{13}C average of $-5^o/oo$ [PDB] has been subjected to an isotopic disproportionation, resulting in the formation of a partial reservoir which is isotopically lighter ($-25^o/oo$) and of another partial reservoir which is isotopically heavier ($\pm 0^o/oo$) than the crustal average, the difference $\Delta\delta$ between these partial boxes always being close to $25^o/oo$ (cf. Degens, 1969; Broecker, 1970; Eichmann and Schidlowski, 1975). With this fractionation value about fixed, and the average isotopic composition of the crustal box as a whole being fixed, too, the relative proportions of these partial reservoirs cannot be varied at will, but are coupled with their respective δ^{13}C averages by a well-defined isotope budget equation.

If we define the ratio R as giving the proportion of organic carbon in the total sedimentary carbon reservoir $[C_{org}/(C_{org} + C_{carb})]$ and express R in terms of the box model of Fig. 1, i.e.,

$$R = B/(B + C),$$ (1)

then we get for $\bar{\delta}_D$ the average $\delta^{13}C$ value of primordial carbon of mantle derivation

$$\bar{\delta}_D = R\bar{\delta}_B + (1 - R)\bar{\delta}_C.$$ (2)

Hence, the average $\delta^{13}C$ value displayed by the stationary sedimentary carbonate reservoir ($\bar{\delta}_C$) reflects a definite value of R, that is, *the relative proportion of organic carbon within the total sedimentary carbon reservoir* [Eq. (1)]. Since the individually measured $\delta^{13}C_{carb}$ values of marine carbonates are basically constant through time, approaching $\bar{\delta}_C$ always within narrow limits, the fossil $\delta^{13}C_{carb}$ record should reflect a similar constancy of R through geologic history. With $\bar{\delta}_D = -5°/oo$, $\bar{\delta}_B = -25°/oo$, $\bar{\delta}_C = {}^+_0°/oo$ [PDB], and reservoir D=1, we would get R=0.2, indicating that organic carbon makes up about one-fifth (or 20%) of the total sedimentary carbon reservoir (cf. Fig. 2).

During the early 1970's, several authors (Schopf et al., 1971; Perry and Tan, 1972; Oehler et al., 1972; Nagy et al., 1974) have submitted isotope values of sedimentary carbon from Precambrian sediments, the quantitatively most important contribution of some 430 values having been made by our group (Eichmann and Schidlowski, 1974, 1975; Schidlowski et al., 1975, 1976a). Figure 3 shows the $\delta^{13}C$ values of 58 coexisting Precambrian C_{org} - C_{carb} pairs as a function of time, covering the whole Precambrian back to almost the start of the sedimentary record. With the exception of a slight shift of some 2.5°/oo toward the positive field in the case of the Late Precambrian samples, the δ values of both C_{org} and C_{carb} are fairly "modern," the overall $\Delta\delta = \delta^{13}C_{carb} - \delta^{13}C_{org}$ for the whole suite being 25.6°/oo.

Figure 4 gives a synopsis of the carbon isotope geochemistry of substantially unaltered sedimentary carbonates through geologic time. With few exceptions (of which the most conspicuous one has, meanwhile, been shown to reflect a special sedimentary environment, cf. Schidlowski et al., 1976b), the $\delta^{13}C$ averages of the carbonate groups keep relatively close to the zero per mil line. Special mention should be made of the relatively narrow standard deviation displayed by 71 samples of Transvaal Dolomite (-0.9 ${}^+_- 1.0°/oo$ vs PDB; Fig. 4, group 5A), which represents the oldest well-preserved shelf carbonate facies of the geologic record. With an overall $\delta^{13}C_{carb}$ mean of +0.4 ${}^+_- 2.7°/oo$ [PDB], the isotopic composition of Precambrian sedimentary carbonates does not differ very much from the average of -0.1 ${}^+_- 2.6°/oo$ [PDB] reported by Keith and Weber (1964) for some 320 Phanerozoic limestones. In terms of the ^{13}C mass balance equation [Eq.(2)], *these rather constant Precambrian*

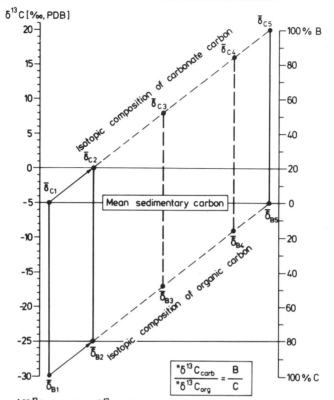

Figure 2

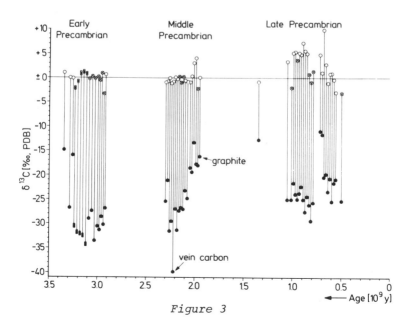

Figure 3

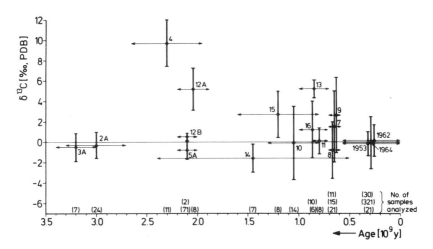

Fig. 4 δ ^{13}C values of substantially unaltered sedimentary carbonates as a function of geologic time. Vertical bars represent standard deviation; horizontal arrows represent possible time range of carbonate groups investigated; numbers refer to tables in original publication (Schidlowski et al., 1975) listing values of original samples. Averages shown are for Phanerozoic carbonates according to Craig (1953), Degens and Epstein (1962), and Keith and Weber (1964).

Fig. 2 Graphic synopsis of constraints imposed by the overall terrestrial ^{13}C mass balance equation [Eq.(2)] on the relative proportions of organic carbon (reservoir B) and carbonate carbon (reservoir C) within the total sedimentary carbon reservoir (vertical bars represent average isotope fractionation $\Delta\delta \cong 25^{o}/oo$ between B and C). With the mean sedimentary $\delta^{13}C_{carb}$ and $\delta^{13}C_{org}$ values of the present reservoir lying at zero and $-25^{o}/oo$, respectively (cf. bar, $\bar{\delta}_{C2}$--$\bar{\delta}_{B2}$), the total sedimentary carbon inventory consists of 20% organic carbon and 80% carbonate carbon.

Fig. 3 δ ^{13}C values of 58 coexisting Precambrian C_{org} - C_{carb} pairs plotted vs geological age (black circles: organic carbon; white circles: carbonate carbon; crossed circles: stromatolitic carbonates). Five Early Precambrian values represented as rectangles have been adopted from Schopf et al. (1971). The average isotope fractionation between C_{org} and C_{carb} for the whole suite is $\Delta\delta = 25.6\%$. From Eichmann and Schidlowski, 1975.

$\delta^{13}C_{carb}$ values lying close to zero per mil would imply that
organic carbon has always made up about one fifth (R $\simeq$ 0.2) of
the stationary sedimentary carbon reservoir since some 3.3 x 10^9
years ago.

IV STATIONARY SEDIMENTARY MASS AND EVOLUTION OF PHOTOSYNTHETIC
OXYGEN

Having thus obtained a reasonable approximation for the
relative proportion of C_{org} within the total sedimentary carbon
reservoir through geologic time, a conversion of R $\simeq$ 0.2 into
an absolute figure would be a necessary prerequisite for calcu-
lating the stoichiometric oxygen equivalent, i.e., the terrestrial oxygen budget of photosynthetic derivation. For this purpose, we have to resort to a second geochemical trick: We have
to introduce R into currently accepted models for the accumulation of the stationary sedimentary mass as a function of time.
The principal reasoning may be outlined as follows.
    Terrestrial sediments can be visualized as being products
of a giant acid titration process taking place on a global
scale between primary igneous rocks and acid volatiles released
from the Earth's interior (Siever, 1968; Garrels and Mackenzie,
1971; Li, 1972). Since a so-called "average igneous rock" should
have always reacted with a mixture of acid volatiles of about
constant composition, the average chemical composition of the
sedimentary shell as a whole should have been approximately constant with time. According to this principle of "chemical uni-
formitarianism" (Garrels and Mackenzie, 1969, 1971), the stationary sedimentary mass may be expected to have always contained
its present total carbon content of roughly 5% (Ronov, 1968),
and our carbon isotope data would indicate that about 20% of
this was always organic carbon from about 3.3 x 10^9 years ago.
    On the other hand, the stationary sedimentary mass--inclu-
sive of the sedimentary carbon reservoir--must have been smaller
some 3 x 10^9 years ago, since degassing of the Earth (which
furnishes the acids for the titration process) is a function of
time. Various lines of evidence as well as theoretical considerations (cf. Birch, 1965; Walker et al., 1970; Fanale, 1971)
lead us to infer that degassing rates must have been very high
immediately after the Earth's formation and that they have ex-
ponentially decreased afterward. Quantitatively, the degassing
process may be best described by the introduction of a degassing
halftime or, respectively, a degassing constant. Recently, Li
(1972) has proposed a value of λ = 1.16 x 10^{-9} year $^{-1}$ as a
"best fit" estimate for the overall terrestrial degassing constant. With this value at hand, the stationary sedimentary
mass M existing at time t may be expressed, in good approxima-

tion, as a fraction of the present mass, $M_{tp} \simeq 2.4 \times 10^{24}$ g, with

$$M_t \simeq M_{tp} (1 - e^{-\lambda t}).$$ (3)

With about 3% of this stationary sedimentary mass M_t being always carbon and the fossil $\delta^{13}C_{carb}$ record indicating that one-fifth of this total carbon was always C_{org}, we may readily calculate both the sedimentary C_{org} reservoir and its stoichiometric oxygen equivalent for any time t covered by the sedimentary record:

$$(C_{org})_t \simeq 0.2 \ \epsilon M_{tp}(1 - e^{-\lambda t})$$ (4)
$$(O_2)_t \simeq 0.53 \ \epsilon M_{tp}(1 - e^{-\lambda t})$$ (5)

(ϵ is the average total carbon content of the sedimentary shell, 0.53 is a stoichiometric conversion factor, tp = 4.5×10^9 years, the age of the Earth).

Figure 5 gives a graphic representation of Eq. (5), assuming 0.53 ϵMtp = $(O_2)_{tp}$ = 1[i.e., all values $(O_2)_t$ are expressed as fractions of the present oxygen reservoir $(O_2)_{tp}$]. According to this function, a stationary total oxygen budget of about 80% of the present one must have already existed very close to the start of the sedimentary record. If the partitioning of this total oxygen between the "bound" and the "free" reservoirs were the same as today (with about 5% of the total being stored as molecular oxygen in the atmosphere - ocean system, see budget column on right side), then the partial pressure of oxygen in the ancient atmosphere should have also amounted to about 80% of the present value (that is, we should have had a pO_2 which presently prevails in a altitude of some 2000 m). This conclusion is, however, incompatible with the paleontological and geological evidence listed in the lower part of Fig. 5. Therefore, we must necessarily assume that, during the early history of the Earth, an almost modern total oxygen reservoir of 80% of the present and more was coupled with negligible O_2 concentrations in the atmosphere. The stationary sedimentary C_{org} reservoir as derived from our model [Eq.(4)] is, accordingly, just a measure of the gross amount photosynthetic oxygen released, withholding all information as to how this oxygen was partitioned between the principal geochemical reservoirs.

It should be noted in this context that the model is based on the premise that the reduced carbon within the sedimentary shell principally owes its origin to the most advanced form of photosynthesis, in which water plays the part of the hydrogen donor and one O_2 molecule is released for every carbon atom incorporated in organic matter. However, as we approach the *dawn of sedimentary history, both nonoxygen-releasing bacterial photosynthesis and abiogenic production of organic matter might become noteworthy contributors of sedimentary organics, although the available paleobiological evidence (cf. Schopf, 1972) seems to favor a very early start of oxygen-producing photosynthetic activity. We would, therefore, not exclude at this stage that a refinement of the basic formalism of Eq. (5) (by introducing an additional co-efficient making allowance for reduced carbon from other sources) might become necessary in due course.*

V. PALEOENVIRONMENTAL IMPLICATIONS OF THE MODEL

The assumed disproportionation between the bound and the free oxygen reservoirs during the Early and Middle Precambrian would imply that the ancient oxygen cycle must have been practically "short-circuited," with all photosynthetic oxygen produced being instantaneously sequestered by the crust as a result of very effective O_2-consuming reactions. In principle, there would be no difficulties in storing the 5% of the total oxygen budget presently contained in the atmosphere - ocean system within the bound reservoir as well, since the reducing capacity of the crust - mantle system is by no means exhausted at the present level of sedimentary O_2 fixation. At present, less than half of the sulfur contained in the ocean - sediment system exists as sulfate sulfur (Li, 1972), the balance being sulfide sulfur, constituting a potential reductant. Furthermore, the Fe^{2+} content of primary rocks exposed at the Earth's surface is likely to provide an almost inexhaustible reducing reserve.

We have tentatively proposed (Schidlowski et al., 1975) that the process of banded ironstone formation characteristic of Early and Middle Precambrian times could have provided such oxygen-absorbing reactions of high efficiency. As had already been previously assumed (MacGregor, 1927; Cloud, 1968, 1973), the ancient seas were likely to have acted as accumulator systems for dissolved bivalent iron ions washed away from the continents as a result of an anoxygenic weathering cycle. The hydrated ions within the ancient seas must have constituted ideal oxygen acceptors when the contemporary autotrophs released molecular oxygen to their environment. With the biological cycle submerged below ocean levels during the Precambrian, the oxygen-scavenging activity of the ferrous ions should have been extremely effective. Only after the ancient oceans had been swept free of dissolved bivalent iron (which came to be precipitated as ferric iron in banded iron deposits) could free oxygen then have accumulated in the seas, and consequently, the atmosphere.

As particularly stressed by Cloud (1968, 1973), the incipient buildup of an atmospheric oxygen pressure is heralded by the appearance of the oldest continental red beds between 1.8 and 2.0 x 10^9 years ago (cf. Fig. 5). Because of the sluggish-

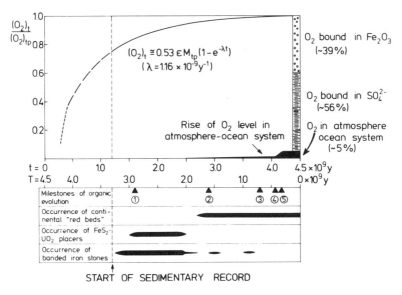

Fig. 5 *Increase of the stationary terrestrial oxygen reservoir of photosynthetic origin as a function of time, with reservoirs existing at times* t *expressed as fractions of the present
reservior* $(O_2)_{tp}$. *The column on the right side shows partitioning
of total oxygen between the bound and the free reservoir in the
present budget according to Li (1972) (note that molecular oxygen makes up only 5% of the present oxygen reservoir). Rise of
the free oxygen level in the atmosphere - ocean system is tentatively inferred from geological and paleontological evidence
represented below. Milestones of organic evolution indicated are
(1) appearance of oldest algal bioherms (photoautotrophic blue-
green algae); (2) appearance of eucaryotic biota; (3) appearance
of oldest eumetazoan faunas; (4) life conquers continents
(Upper Silurian); (5) appearance of exuberant continental
floras (Upper Carboniferous).*

ness of oxidation weathering on the continents (where rocks are
being slowly decomposed and their reducing constituents thus made
available to oxygen attack), a new stationary state must have
finally been established at a markedly increased atmospheric
pO_2 level at which the new sink function ultimately matched the
rate of oxygen production. The oxygen burden of the present
atmosphere must, therefore, be looked upon as being due to the
failure of thermodynamics to exercise strict control over the
redox balance of the atmosphere - hydrosphere - crust system
during operation of the present weathering cycle, *with dynamic
equilibrium between source and sink being attained only at a
considerable* pO_2 *in the free reservoir.*

Acknowledgments

*This chapter is a digest of a previous publication by M. Schid-
lowski, R. Eichmann, and C. E. Junge (1975, Precambrian Res.
, 2;
1) based on an Invited Lecture delivered by the senior author at
the College Park Colloquia on Chemical Evolution (University of
Maryland, 1975). The work as a whole was performed as part of
the SFB 73, receiving partial funding from the Deutsche Forsch-
ungsgemeinschaft.*

REFERENCES

Birch. F. 1965. *Geol. Soc. Amer Bull.*, *76*:133.
Bond, G., Wilson, J. F., and Winnall, N. J. 1973. *Nature (London)*, *244*:275.
Broda, E. 1975. *The Evolution of Bioenergetic Processes*, p. 220. Pergamon, Oxford.
Broecker, W. 1970. *J. Geophys. Res.*, *75*:3553.
Broecker, W. 1973. In G. M. Woodwell and E. V. Pecan (eds.), *Carbon and the Biosphere*, pp. 32-50. U. S. Atomic Energy Commission, Washington, D. C.
Cloud, P. E. 1968. *Science*, *160*:729.
Cloud, P. E. 1973. *Econ. Geol.*, *68*:1135.
Craig, H. 1953. *Geochim. Cosmochim. Acta*, *3*:53.
Degens, E. 1969. In G. Eglington and M. T. J. Murphy (eds.), *Organic Geochemistry*, pp. 304-329. Springer, Berlin.
Degens, E., and Epstein, S. 1962. *Amer. Assoc. Pet. Geol. Bull.*, *46*::534.
Deines, P., and Gold, D. P. 1973. *Geochim. Cosmochim. Acta*, *37*: 1709.
Eichmann, R., and Schidlowski, M. 1974. *Naturwissenschaften*, *61*:449.
Eichmann, R., and Schidlowski, M. 1975. *Geochim. Cosmochim. Acta*, *39*:585.
Fanale, F. P. 1971. *Chem. Geol.* *8*:79.
Garrels, R. M., and Mackenzie, F. T. 1969. *Science*, *163*:570.
Garrels, R. M., and Mackenzie, F. T. 1971. *Evolution of Sedimentary Rocks*, p. 397. Norton, New York.
Garrels, R. M., and Mackenzie, F. T. 1972. *Marine Chem.*, *1*:27.
Hawkesworth, C. J., Moorbath, S., O'Nions, R. K., and Wilson, J. F. 1975. *Earth Planet. Sci. Lett.* *25*:251.
Holland, H. D. 1976. *The Chemistry and Chemical Evolution of the Atmosphere and Oceans*, in press.
Junge, C. E., Schidlowski, M., Eichmann, R., and Pietrek, H. 1975. *J. Geophys. Res.* *80*:4542.
Keith, M. L., and Weber, J. M. 1964. *Geochim. Cosmochim. Acta*, *28*:1787.
Li, Y. H. 1972. *Amer. J. Sci.*, *272*:119.

MacGregor, A. M. 1927. *S. Afr. J. Sci.*, *24*:155.

MacGregor, A. M. 1940. *Trans. Geol. Soc. S. Afr. 43*:9.

Nagy, B., Kunen, S. M. Zumberge, J. E., Long, A., Moore, C. B., Lewis, C. F., Anhaeusser, C. R., and Pretorius, D. A. 1974. *Precambrian Res.*, *1*:43.

Oehler, D. Z., Schopf, J. W., and Kvenvolden, K. A. 1972. *Science*, *175*:1246.

Perry, E. C., and Tan, F. C. 1972. *Geol. Soc. Amer. Bull.*, *83*: 647.

Rabinowitch, E. I. 1945. *Photosynthesis and Related Processes.* Interscience, New York.

Ronov, A. B. 1968. *Sedimentology, 10*:25.

Schidlowski, M., Eichmann, R., and Junge, C. E. 1975. *Precambrian Res.*, *2*:1.

Schidlowski, M., Eichmann, R., and Fiebiger, W. 1976a. *N. Jb. Miner. Mh.*, *1976*:344.

Schidlowski, M., Eichmann, R., and Junge, C. E. 1976b. *Geochim. Cosmochim. Acta, 40*:449.

Schopf, J. W. 1972. In C. Ponnamperuma(ed.), *Exobiology*, pp. 16 - 61. North-Holland, Amsterdam.

Schopf, J. W., Oehler, D. Z., Horodyski, R. J., and Kvenvolden, K. A. 1971. *J. Paleontol.*, *45*:477.

Siever, R. 1968. *Sedimentology, 11*:5.

Sillen, G. 1961. In M. Sears (ed.), *Oceanography*, pp. 549 - 581. *Amer. Assoc. Adv. Sci. Publ.*, *67*:Washington, D.C.

Walker, J. C. G., Turekian, K. K., and Hunten, D. M. 1970. *J. Geophys. Res.*, *75*:3558.

Welte, D. 1970. *Naturwissenschaften, 57*:17.

11

EVIDENCES OF ARCHEAN LIFE

J. WILLIAM SCHOPF
University of California

Eight principle types of data, including both (possible) "chemical evidence" and (possible) "morphological evidence," have been suggested by various workers as being indicative of the existence of Archean biological activity. Although most such data are consistent with this interpretation, few, if any, are wholly compelling. A brief summary follows.

I. CHEMICAL EVIDENCE

A. *Extractable Organic Matter*

Amino acids, fatty acids, porphyrins, *n*-alkanes, and isoprenoid hydrocarbons have been detected in solvent extracts of Archean sediments (Kvenvolden, 1972). No data are available to establish firmly that these materials date from the time of original, Archean, sedimentation.

B. *Kerogenous carbon*

Acid- and solvent-resistant particulate organic matter (kerogen), often occurring in fine sedimentary layers and thus demonstrably syngenetic with sediment formation, occurs in many Archean deposits (Kvenvolden, 1972). The source(s) of this kerogen---whether biological, abiological, or of combined origins---has not been established.

C. *Carbon Isotopic Abundances*

A substantial number of analyses has been made of carbon isotopic abundances in Precambrian carbonates (Schidlowski *et al.*, 1975), in Precambrian organic matter (Oehler *et al.*, 1972), and in coexisting organic carbon - carbonate pairs in Precambrian sediments (Eichmann and Schidlowski, 1975). Although questions remain regarding the probable mass distribution of carbon in the various Archean reservoirs and the degree of isotopic fractionation that may have occurred during abiotic syntheses of Archean organic matter, these data are at least suggestive of the existence of Archean life.

D. *Oxygen in Banded Iron Formations*

It has often been speculated that the oxygen required for oxidation of iron minerals occurring in Precambrian banded iron formations may have been of biological origin; thus, banded iron formations (dating back to about 3750 million years) have been regarded as indirect evidence of biological activity (Cloud, 1974). Alternatively, such oxygen may have been solely nonbiological in origin; if so, iron formations represent evidence consistent with, but not necessarily indicative of, the presence of oxygen-producing photoautotrophs (or, indeed, of the necessary existence of any type of biological activity) (Schopf, 1975).

II. MORPHOLOGICAL EVIDENCE

A. *Stromatolites*

Laminated, carbonate stromatolites are now known from four Archean horizons: from the Steep Rock Lake Series in the Atikokan region of Ontario, Canada--?2600 million years old (Hofmann, 1971); from the Yellowknife Supergroup at Snofield Lake, Northwest Territories, Canada--more than 2600 million years old (Henderson, 1975); and from two horizons in the Bulawayan Group--both 2800 to 2500 million years old (Hawkesworth *et al.*, 1975)--one near Bulawayo, Rhodesia (Schopf *et al.*, 1971) and the other near Belingwe, Rhodesia (E. G. Nisbet, personal communication, 1975). These forms are essentially indistinguishable from younger stromatolites that are of demonstrable biogenicity. Although (like virtually all carbonate stromatolites) they do not contain structurally preserved microfossils, there seems no reason to suspect that they could have resulted solely from inorganic, accretionary processes. They are highly suggestive of Archean biological activity.

B. *Ultramicrofossils*

Although a rather large number of microstructures, regarded as Precambrian fossils, has been reported from preparations of Archean sediments studied by transmission electron microscopy, most, and perhaps all, of these structures now appear to be artifacts, and assuredly nonbiological, or to be recent, and primarily bacterial, contaminants (for detailed discussion see Schopf, 1975).

C. *Filamentous Microfossils*

A number of workers have reported the occurrence of "filaments," "filamentous bodies," "fibrillar structures," etc., from sediments of Archean age. Although the existence of well-laminated Archean stromatolites provides presumptive evidence for the early appearance of the filamentous habit, none of the reports of Archean filamentous microfossils published to date seems compelling (Schopf, 1975).

D. *Spheroidal Microfossils*

In recent years, numerous workers have reported the occurrence of microscopic organic spheroids in Archean sediments (Schopf, 1975); data have been published summarizing the size distribution of more than 1300 such bodies from the Swaziland Supergroup of South Africa (Fig. 1). Statistical analyses have shown, however, that populations of these spheroids bear more resemblance to those of the carbonaceous "organized elements" of the Orgueil meteorite than they do to populations of modern unicellular algae or of assured Precambrial fossils (Schopf, 1976). Moreover, as shown in Fig. 1, the size ranges reported for these spheroids are much broader than, and thus quite unlike those known for assured unicellular microfossils preserved in similar facies of younger Precambrian age. In addition, they differ distinctly from the size range exhibited by modern procaryotic algae (Fig. 1). At present, it appears likely that the carbonaceous spheroids reported from Archean sediments are at least partly, and perhaps wholly, of nonbiological origin.

REFERENCES

Cloud, P. E. 1974. *Am. Scientist, 62*:54.
Hawkesworth, C. J., Moorbath, S., o'Nions, R. K., and Wilson, J. F. 1975. *Earth Planet Sci. Lett. 25*:251.
Henderson, J. B. 1975. *Geol. Surv. Canad.* Paper 74-1, Pt.A, 325.
Hofmann, H. J. 1971. *Geol. Surv. Canad. Bull. 189*:146pp.
Kvenvolden, K. A. 1972. *24th Intern Geol. Congr., Sect. 1,* Montreal,

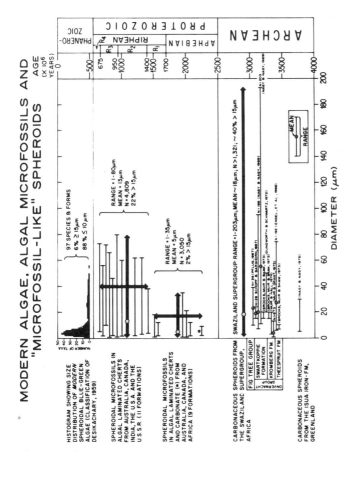

FIG. 1. Modern Algae, Algal Microfossils and Microfossil-Like Spheroids.

Eichmann, R., and Schidlowski, M. 1975. *Geochim. Cosmochim. Acta*
   39:588.
Kvenvolden, K. A.1972. *24th Intern. Congr. Geol. Sect. 1*,
   Montreal: 31.
Oehler, D. Z. Schopf, J. W., and Kvenvolden, K. A. 1972. *Science*,
   175:]246.
Schlidowski, M., Eichmann, R., and Junge, C. E. 1975. *Precambrian
   Res.*, *2*:1.
Schopf, J. W. 1975. *Ann Rev. Earth Planet. Sci.*, *3:*213.
Schopf, J. W. 1976. *Origins of Life*, *7:*19.
Schopf, J. W., Oehler, D. Z., Horodyski, R. J., and Kvenvolden,
   K. A. 1971. *J. Paleontol.*, *45:* 477.

12

EARLIEST EVIDENCE OF FOSSIL EUCARYOTES

J. WILLIAM SCHOPF
University of California

Among the dozen or so most important evolutionary events to have occurred during the history of life on this planet, the origin of the eucaryotic cell type stands out as having had special impact, an event that changed the rate and altered the course of early evolutionary development. Although paleobiologic data can be expected to provide only limited insight into the *mode* of origin of the eucaryotes, evidence indicative of at least a minimum age for the *timing* of this event--the earliest appearance of distinctive eucaryotic characteristics in fossil organisms--should be preserved in the fossil record. As is typical of paleobiology in general, however, assessment of such evidence is a somewhat subjective matter; evidence regarded by one worker as "compelling" or even "conclusive" may be regarded by others as "consistent" but merely "suggestive." The problem is further compounded by the nature of the procaryotic to eucaryotic transition--regardless of mode of origin, the earliest eucaryotes seem certain to have been simple, microscopic unicells that differed principally from their procaryotic progenitors only in intracellular organization. They cannot reasonably be expected to have exhibited traits widespread in more complex eucaryotes (e.g., tissue and advanced sporangial formation, true branching, polarity of organization, megascopic size, etc.), traits that might seem clearly indicative of eucaryotic affinities. The problem thus becomes one of defining those potentially preservable traits that can be expected to have typified primitive eucaryotes and determining the time of first appearance of such traits in the fossil record.

Results of such analyses, based on studies of Precambrian microfossils occurring in about 24 algal-laminated cherts (Schopf, 1975a) and in more than 100 carbonaceous shales (Timofeev, 1974), lead me to conclude that nucleated organisms were extant prior to 850 ± 100 million years ago (Schopf and Blacic, 1971; Walter, 1972) and that the lineage may well have been established as early as 1400 ± 100 million years ago. A brief summary of the evidence (also see Schopf and Oehler, 1976):

(1) Spheroidal organic bodies, interpreted as being remnants of intracellular organelles (and thus as being indicative of eucaryotic organization), occur in algal microfossils about 850 ± 100 million years old (Schopf, 1975a, Schopf and Blacic, 1971; Walter, 1972). These organelle-like bodies, comparable in size and shape to nuclei and pyrenoids in modern eucaryotes of similar cell dimensions, can be differentiated by both optical microscopy and transmission electron miscroscopy from remnants of degraded protoplasm that occasionally occur in the same cells (Schopf and Oehler, 1976).

(2) Filamentous microorganisms, exhibiting features suggestive of fungal (and thus eucaryotic) affinities, are known from sediments about 850 ± 100 years old (Schopf and Blacic, 1971; Walter, 1972). A tetrahedral tetrad of algal unicells, similarly suggestive of eucaryotic organization, occurs in the same deposit.

(3) Irregularly septate, highly branched filaments of large diameter (15 - 60 μm), similar in organization to siphonaceous green or golden green (vaucheriacean?) eucaryotic algae, are known from cherts about 1300 and 1000 million years old (Schopf, 1975a).

(4) Studies have been made of the size distribution (range and mean cell size) of about 7000 simple spheroidal microfossils measured in petrographic thin sections of algal-laminated sediments from 20 formations from Australia, Canada, India, Africa, the U.S.A., and the U.S.S.R. (Schopf, 1975b). All assemblages older than about 1400 ± 100 million years contain small cells that exhibit distribution patterns comparable to those of modern procaryotic algal taxa. In contrast, all younger assemblages contain relatively large cells (40 μm), and many have mean cell sizes (∿17 m) larger than all but a few taxa of modern coccoid cyanophytes (∿10% > 15 μm), and most contain cells (60 - 80 μm) larger than any known procaryotes, but of a size exhibited by modern eucaryotes. The distribution patterns exhibited by assemblages younger than about 1400 ± 100 million years old suggest that they are composed of a mix of procaryotic and eucaryotic taxa.

(5) Very large unicells (50 - 300 μm), up to six times larger than any known procaryotes, occur in shales of about 850 ± 50 million years of age (Timofeev, 1974). Microfossils extracted from shales also exhibit a rather abrupt increase in

cell size at 1400 $\pm$ 100 million years ago and the occurrence of
a marked nearly fivefold increase in generic diversity between
about 1400 and 1100 million years ago, which was followed by a
period of quiescence (featuring an increase in generic diversity
of only 4%) until the close of the Precambrian (Timofeev, 1974).

REFERENCES

Schopf, J. W. 1975a. *Ann. Rev. Earth Planet Sci.*, 3:213.
Schopf, J. W. 1975b. *Chemical Evolution of the Precambrian* (Abstr.)
     College Park Colloquia on Chemical Evolution, p. 30.
Schopf, J. W., and Blacic, J. M. 1971. *J. Paleontol.*, 45:925.
Schopf, J. W., and Oehler, D. Z. 1976 *Science, 193*:47.
Timofeev, B. V. 1974. *Proc. 3rd. Int. Conf. Palynol. Nauka, 7.*
Walter, M. R. 1972. *Palaeontol. Assoc. London Spec. Pap., 11.*

13

PALEOBIOLOGY OF STROMATOLITES

STANLEY M. AWRAMIK

University of California, Santa Barbara

Most of our understanding of evolutionary events before the appearance of Metaphyta and Metazoa is based on the detailed study of microfossils found in cherts. These cherts, though rare in pre-Phanerozoic sediments, are yielding a rich and diverse microbiota from which a story of increased diversity and morphologic complexity with time is unfolding. Very little has been added to this understanding using the other great class of pre-Phanerozoic fossils--the stromatolites. Stromatolites are organosedimentary macrostructures produced by sediment trapping, binding, and precipitation activity of microorganisms--predominantly blue-green algae.

The vast majority of stromatolites found in the pre-Phanerozoic is found preserved as carbonates, rarely as cherts. Consequently, due to the low preservation potential of nonskeletal microorganisms in carbonates, most stromatolites are barren of microfossils. What kind of biological information can we obtain from stromatolites that will add to our understanding of the pre-Phanerozoic?

Stromatolites first appear in the Archean. Four horizons, all in carbonates, are known to date: two in Canada and two in Rhodesia. Assuming the structures are stromatolites, certain statements concerning the level of evolution in the early pre-Phanerozoic can be made:

(1) Microorganisms evolved the ability to live benthically early in their history.

(2) They evolved either through motility or rapid growth, the ability to remain at or near the sediment - fluid interface keeping pace with sedimentation rates,

---

[*]Contribution No. 63 of the Biogeology Clean Lab.

(3)   By analogy with younger and presently forming stro-
matolites: (a) they were probably cyanophytes; (b) a community
of different microorganisms may have been involved including
blue-green algal producers and bacterial decomposers; (c)
the atmosphere, at least locally, was able to support carbo-
nate precipitation and accumulation concomitant with photo-
synthetic activity; and (d) the stromatolitic habit must have
had great selective advantages for microorganisms if we look
at the richness of the stromatolite record in the Proterozoic.

By far the greatest information on the paleobiology of
pre-Phanerozoic stromatolites is being obtained from detailed
studies of the microfossiliferous stromatolitic cherts of the
Gunflint Iron Formation in Canada, some 2000 million years old.
The stromatolites are dominated by two microfossil morphotypes:
(a) the filamentous Gunflintia minuta of presumed blue-
green algal affinity; and (b) Huroniospora sp., a presumed
coccoid blue-green.   I find a change in the proportion of diff-
erent microfossil morphotypes in stromatolites of differing
morphology.   Thus, by Gunflint time, complex communities inter-
acted with the physical, and possibly chemical, environment
giving rise to biologically well-defined stromatolite morpholo-
gies.   Observations on the Gunflint, coupled with the prospect
that certain stromatolite morphologies may be useful in pre-
Phanerozoic biostratigraphy, open a new door for paleobiological
research: the biological significance of stromatolite morpho-
logy and its relationship to microbial evolution in the pre-
Phanerozoic.

I.   INTRODUCTION

In the quest for a greater understanding of the pre-Phanerozoic history of life, biogeologists have utilized three major
groups or types of fossil evidence: (a) microorganisms preserved in cherts and, to a lesser extent, in shales and carbonates; (b) stromatolites; and (c) the organic geochemistry of
sedimentary rocks.
     Microfossils and stromatolites at present provide the
most direct and potentially paleobiologically significant data
with which to interpret the early evolutionary history of life.
Chemical fossils found in pre-Phanerozoic rocks are subject
to questions concerning their indigenous nature in the rock,
whether biological, abiological, or of combined origins, and
what the original organic precursors were (McKirdy, 1974; Schopf,
1975b).
     Most of our understanding of evolutionary events before
the rise of multicellular organisms is based on detailed studies of microorganisms found preserved in various sedimentary

rock types, but principally in cherts. Localities with diverse,
well-preserved microbiotas are rare, with only 20 recognized
(Schopf, 1975a). These localities are yielding rich and diverse microbial assemblages from which a picture of increasing
diversity and morphologic complexity with time is unfolding.
Stromatolites, though very conspicuous and abundant in Proterozoic carbonates, have contributed little to this evolutionary
picture.

The paleobiology of stromatolites can be better understood
within a conceptual framework based on our knowledge of Recent
algal mats and the biological processes associated with mats.
Much of the information we have on the paleobiology of stromatolites has been based on observational data noting stromatolite size, shape, microstructure, distribution in time and
space, and comparison to Recent "analogues." Little attention
has been devoted to the fundamental question of what does the
sheer presence of a stromatolite mean paleobiologically? What
assumptions can be made and what conclusions drawn? Such a
conceptual framework can yield potentially valuable data with
which to better understand life in the pre-Phanerozoic. Because stromatolites are so abundant in pre-Phanerozoic rocks,
this line of questioning is producing interesting and potentially significant insight into the world of stromatolites. I
utilize this approach fully realizing that my premises are
based on our conventional wisdom concerning algal mats and their
accreted products, the stromatolites.

II. THE STROMATOLITES

I view stromatolites as organosedimentary structures produced by sediment trapping, binding, and/or precipitation resulting from the metabolic activity and growth of microorganisms, primarily blue-green algae. The term stromatolite is
generally reserved for laminated structures exhibiting vertical
relief above the substrate, forming domes or columns.

The vast majority of stromatolites found in the pre-Phanerozoic is preserved as carbonates, rarely as cherts or with
chert within carbonate stromatolitic laminae. Chert, if present,
seldom contains microfossils. Therefore, we have extremely little data on the microorganisms responsible for producing the
majority of stromatolites in the fossil record. Studies on Recent algal mats and stromatolites indicate that benthic nonskeletal photosynthetic organisms dominate; in particular, blue-green
algae (Ginsburg et al., 1954; Monty, 1967; Logan et al., 1974).
Due to the extremely low preservation potential of nonskeletal
organisms in carbonates, stromatolites are essentially barren
of microfossils. Korde (1953) and Vologdin (1955, 1962) have

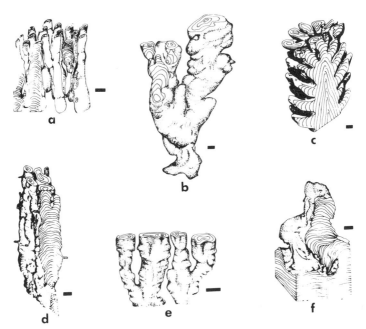

Fig. 1.   Six graphically reconstructed stromatolites.   (a)
Gymnosolen Steinmann (in Bertrand-Sarjati, 1972, p. 143), (b)
Tungussia erecta (Walter, 1972, p. 172), (c) Jacutophyton
Shapovalova (in Serebryakov, 1975, p. 106), (d) Kulparia kulpa-
rensis (Preiss, 1973, p. 112, (e) Inseria tjumosi Krylov (in
Raaben, 1969, p. 78), (f) Baicalia burra (Preiss, 1972, p. 83).
Bar equals 1 cm.

identified what they interpret as microorganisms preserved within the laminae of carbonate stromatolites from the pre-Phanerozoic through the Cambrian of the U. S. S. R. and China. Serious
doubt exists as to whether all the structures described and illustrated by Korde and Vologdin are the actual cellular remains
of algae, degraded organic matter, and/or diagenetically and
mineralogically altered artifacts (Korolyuk, 1960, 1963; Semikhatov et al., 1963; Krylov, 1963; Raaben and Komar, 1964; Kirichenko, 1964; Walter, 1972).

The most compelling evidence for cellular remains within
stromatolites is found in microfossiliferous stromatolitic
cherts. I recognize 11 stromatolitic chert localities for which
adequate published data on the type of stromatolite present and
associated microfossils taxonomically are available (Table 1).

Microfossil diversity is low at most of these localities,
which may reflect the following possibilities: (a) preliminary
nature of the report; (b) a low diversity in the ancient stro-

TABLE 1

	GEOLOGIC UNIT	LOCALITY	AGE m.y.	COLUMNAR	STRATIFORM to DOMAL	ONCOLITE	BACTERIA	CYANOPHYTA NOSTOCALES	CHROOCOCCALES	CHLOROPHYTA	CHLOROPHYTA or RHODOPHYTA	CHLOROPHYTA or CHRYSOPHYTA	PYRROPHYTA	EUMYCOPHYTA	UNKNOWN AFFINITY	REFERENCES
RIPHEAN	Bitter Springs Fm.	Northern Territories, Austr.	~900		X		3	27	11	2	4		1	2	2	Schopf 1968; Schopf and Blacic 1971
	Skillogale Dol.	South Australia	~1000		X		?1	?4	2	2	1					Schopf and Barghoorn 1969; Schopf and Fairchild 1973; Knoll, et al 1975
	Dismal Lakes Grp.	NWT, Canada	~1200		X			4	1							Donaldson and Delaney 1975
	Beck Springs Dol.	SE California	~1300		X	X	?1	3	1	2		?4				Cloud et al 1969; Licari, in press
	Bungle Bungle Dol.	Western Australia	~1500		X			2	?2	1	2					Diver 1974
	Amelia Dol.	Northern Territories, Austr.	~1600		X			1	5						1	Muir 1974 and in press
	Paradise Creek Fm.	Queensland, Austr.	~1650	X	X			2	2							Licari, Cloud and Smith 1969; Licari and Cloud 1972
	Belcher Group	Hudson Bay, Canada	~1800		X		?3	3	3					1		Hofmann and Jackson 1969
1700 m.y. PRE-RIPHEAN	Gunflint Iron Fm.	Ontario, Canada	~2000	X	X		?2	3	?3						4	Barghoorn and Tyler 1965; Awramik and Barghoorn 1975 and in prep.
	Kasagalik Fm.	Hudson Bay, Canada	~2000		X		1	2	2							Hofmann 1973, 1974; Hofmann and Golubic, in press
	Transvaal Dol.	South Africa	~2300	X			2	2	1							Nagy 1974; MacGregor, et al 1974

115

matolite community; (c) preferential preservation of certain
morphotypes; and/or (d) preferential silicification of certain
regions of an initially vertically differentiated algal community
(see Golubic, 1973, for a discussion of vertically differentiated communities). High diversities approaching 60 different species have been observed in the Recent algal mats from the Per-
sian Gulf (Golubic, personal communication, 1974). However, one
to a few species commonly dominate in Recent mats, with the
additional species observed occupying microenvironments within
the mat (Awramik et al., 1976).

     Clearly, stromatolites, though abundant, supply little
direct cellular data with which to interpret ancient algal mat
microbiotas and their evolution in time.

III. STROMATOLITE BIOSTRATIGRAPHY

     Since about 1960, geologists, first in the U. S. S. R. and
later in countries the world over, have found certain stromatolites and assemblages of stromatolite morphologies to have restricted time ranges within the pre-Phanerozoic (Keller et al.,
1960; Krylov, 1963, 1967; Semikhatov and Komar, 1965; Cloud and
Semikhatov, 1969; Valdiya, 1969; Walter and Preiss, 1972).
These studies have resulted in a subdivision of the last 1000
million years of pre-Cambrian time into four time-stratigraphic units, based on assemblages of distinctive columnar stromatolites. The early work was empirical with no biogeological
significance attached to these time-restricted structures.
I liken these early but reasonably successful attempts using
stromatolites as time markers to William Smith utilizing fossils
in 1797 and 1815 to subdivide rock units in England and Wales.
Both Smith and the early stromatolite workers devoted little
attention to the possible biological significance of the fossils.
Smith did not have the benefit of modern evolutionary theory,
while the pioneering stromatolite biostratigraphers were only
interested in their utility. But both arrived at the same conclusion using fossils--that they worked biostratigraphically.

     The most widely used and successful stromatolite morphology
so far employed to subdivide the pre-Phanerozoic is the columnar
types (Fig.1). Interestingly enough, only three of the localities in Table 1 have microfossils preserved within columnar
stromatolites, and all are near or below the lower boundary of
the time-stratigraphic scheme established using stromatolites.
Pre-Riphean stromatolite biostratigraphy has too little data
(see Butin, 1966; Cloud and Semikhatov, 1969; Walter, 1972) with
which to establish a global scheme comparable to that developed
for the Riphean.

     That certain stromatolites have restricted time ranges
within the pre-Phanerozoic necessitates some biological/ecologi-

TABLE 2

Summary of Gunflint Stomatolite Data

Stromatolite	Algae	Range of Ooids and Clasts Max. Apparent Dia.	Average Max. Dia. Ooids and Clasts
Gruneria biwabikia	coccoid 30-34%	0.1-0.9 mm	0.5 mm
	filaments 66-70%		
Kussiella	coccoid 42-46%	0.3-11 mm	4.0 mm
	filaments 54-58%		
Stratifera	coccoid 65-68%	0.0-1.5 mm	0.8 mm
	filaments 32-35%		
Unnamed stromatolite	coccoid 44-48%	0.2-2.0 mm	1.2 mm
	filaments 52-56%		

cal explanation. Either microbial evolution or atmospheric and
hydrospheric evolution were responsible, or both acting in concert produced these stromatolitic guide fossils. Most likely
both acted in concert.

IV. PALEOBIOLOGY OF STROMATOLITES

Though we have little observational date on the paleomicrobiology of ancient stromatolites, there are several lines of
circumstantial evidence to consider. The most fundamental as-
pect to consider is what kind of paleobiological information
does the sheer presence of a stromatolite hold?
Given any of the nonmicrofossiliferous stromatolites illustrated in Fig. 1, we can infer the following, based on experience with Recent algal mats and stromatolites (see also Awramik *et al.*, in press):

(*1*) Blue-green algae were present,
(*2*) They were benthic through all or part of their life
cycle,
(*3*) The structure was probably produced by filamentous
blue-green algae (Schopf *et al.*, 1971),
(*4*) Few to many species probably were present,
(*5*) Blue-green algae evolved either through motility or
rapid growth, the ability to remain at or near the sediment
fluid interface keeping pace with rates of sedimentation, and/
or mineral precipitation--yet at an optimum level of light,
enabling photosynthesis,
(*6*) The probable evolution of a sheath or envelope for
several selective advantages: (*a*) the mucilage would aid in
trapping and binding sediment, and living within sediment may
have sheltered early blue-greens from harmful radiation; (*b*)
the sheath itself may have provided some protection from harmful
radiation; (*c*) the mucilage could help the microorganisms maintain a benthic existence; and (*d*) mucilage also aids the cohesive nature of algal mats--thus enabling the mat as a whole, and
its algal builders, to withstand varying degrees of turbulence.
(*7*) The community included blue-green algal producers and
bacterial decomposers.

Isolated occurrences of mats and stromatolites produced
in part or dominated by microorganisms other than blue-green
are known; bacteria-dominated mats in siliceous hot springs
(Walter *et al.*, 1972) and bacterially produced laminated man-
ganese nodules from abyssal plains (Monty, 1973a). These environments, though of great interest in terms of the microbiological convergence on the stromatolite-building habit, have
no apparent bearing on the majority of stromatolites which grew

in shallow-water marine environments. Blue-green algae are, and
presumably were, responsible for almost all stromatolites. Indirect evidence for blue-green algae, or at least photosynthetic
activity utilizing CO_2, is provided by the carbonate nature of
most stromatolites. With photosynthesis and the uptake of CO_2,
precipitation of $CaCO_3$ is favored. It may not be fortuitous
then, that the majority of stromatolites are found preserved as
carbonates.

    Given the inferences presented above, we can see that stro-
matolites provide some indirect paleobiological evidence. Stromatolites can also provide a minimum indication of the level of
evolution achieved at the time the rocks in question were deposited. This at first may appear to be a strange statement, but
given a stromatolite, one can say with a relatively high degree
of confidence that blue-green algae were present. This statement,
as presumptive as it may be, has significance when applied to
stromatolites known from the Archean.

    Stromatolite microstructure (that is, the textural details
of the individual lamina) in a broad sense includes: synoptic
profile (Hofmann, 1969b), or relief above a single growth surface;
microcrenulations; grain-to-grain relationships of sediment trapped
or precipitated and other sedimentological characteristics; and
distribution of the organic matter (Gebelein, 1974; Walter, 1972).
Microstructure has been used in stromatolite biostratigraphy to
define certain forms (see Korolyuk, 1960; Semikhatov, 1962). Microstructure appears to be strongly influenced by the algae which
trap, bind, and/or precipitate sediment (Gebelein, 1974). The
degree to which algal communities dominated by one family of cyanophytes actually impart a characteristic microstructure is unclear, contrary to the conclusions of Gebelein (1974). Recent
studies (Golubic and Awramik, personal observations, 1973) show
that oscillatoriaceans, both large with thick sheaths and small
with thin sheaths, can produce microstructurally similar structures
with several millimeters of relief along a single lamina (Fig. 2).

    The degree of taxonomic refinement achieved based on
conclusions regarding stromatolite microstructure is uncertain
at present. The prediagenetic microstructure of Recent stromatolites appears to be biologically controlled. The effect of algae occupying lower levels within the vertically differentiated
community on microstructure imparted by algae at the surface of
the mat is unknown. We do not yet understand postdeposition
bacterial interaction with the accreted sediments and organic
matter, nor their impact on microstructure coupled with diagenetic changes.

    Schopf (1975b) outlined eight principal types of data that
have been suggested to be indicative of Archean biological activity: (a) extractable organic matter; (b) kerogenous carbon; (c)
carbon isotopic abundances; (d) oxygen in banded iron-formations;

a

b

Fig. 2 Pinnacle or tufted mats from Hamelin Pool, Shark Bay, Western Australia. (a) Tufted mat built by Lyngbya, a large, thick sheathed cyanophyte. (b) Mat with tufts built by Schizothrix, a small, thin-sheathed cyanophyte.

(e) stromatolites; (f) "ultramicrofossils"; (g) filamentous "microfossils"; and (h) speroidal "microfossils." Schopf (1975b) states: "Although most such data are consistent with this interpretation, few, if any, are wholly compelling (p. 30)."

The presence of stromatolites"...essentially indistinguishible...(p.31)" from younger stromatolites demonstrably of biogenic origin in rocks of Archean age may comprise the most compelling line of evidence for life in the Archean. Four Archean localities containing laminated domal, pseudocolumnar to columnar carbonate stromatolites are known: two in Canada and two in Rhodesia. Assuming these structures are stromatolites, then concomitant with this assumption, photosynthetic microorganisms evolved during Archean time. Most likely they were blue-greens and filamentous. The other assumptions previously made could also apply. The story, though relying on assumptions and indirect lines of evidence, is consistent with available data (see also Schopf, 1975a,b).

In carbonate rocks of Proterozoic age, stromatolites become increasingly abundant and markedly diverse in form. The morphologic diversity of columnar stromatolites increases from the pre-Riphean through the Riphean, reaching maximum diversity during the Upper Riphean (950 to 675 million years ago) but decreasing in diversity during the Vendian (675 to 570 million years ago) and early Paleozoic (Awramik, 1971a). This increase in diversity, and its timing, correlate well with the increase in diversity of microfossils during the Riphean. · The decrease in diversity during the Vendian and the Cambrian may be related to the origin of grazing and burrowing metazoans (see Awramik, 1971a). The increasing abundance and diversity of stromatolites from about 2 billion years to 700 million years ago has prompted Cloud (1975) to refer to this interval as the age of stromatolites. The preponderance of Proterozoic stromatolites, which presumably grew in shallow marine environments, indicates a great deal of blue-green algal activity and bacterial decomposition.

By Transvaal time, some 1950 to 2300 million years ago, the stromatolitic habit became established in the intertidal region as well as in the shallow subtidal (Button, 1971; Truswell and Ericksson, 1972). This "invasion" of the intertidal necessitated the evolution of additional mechanisms to cope with a periodically exposed environment (Awramik et al., 1976). These mechanisms would evolve to cope with: (a) shorter wavelengths of light; (b) high intensity visible light which, in the presence of O_2, decomposes chlorophyll and carotenoids, leading to potentially lethal epoxide compounds (Krinsky, 1966; Leff and Krinsky, 1967); (c) water loss; and possibly (d) rapidly changing temperatures and salinities. Thus, the intertidal habit required further refinement of an already sophisticated blue-green algal physiology and appeared early in the history of both blue-green algae and stromatolites.

V. GUNFLINT STROMATOLITES

Until now, I have been developing a conceptual framework with which to understand the paleobiology of unmicrofossilferous stromatolites. Though based on presumptive evidence, this picture is consistent with data available.

The greatest amount of information available on the paleobiology of ancient stromatolites comes from the Gunflint Iron Formation, some 2000 million years old, from the north shore of Lake Superior. Much background material has been written on the microbiota (see principally Barghoorn and Tyler, 1965; Cloud, 1965; Awramik and Barghoorn, 1975) and the stromatolites (see Hofmann, 1969a) and need not be repeated here.

The stromatolites of the Gunflint Iron Formation, with their well-preserved and diverse microbiota, provide valuable data about and insight into stromatolite-building communities back

some 2000 million years (Awramik, 1976). The conclusions
deduced from the Gunflint may be applicable, with caution, to
unmicrofossiliferous stromatolites. It is the only locality
yet known that has microfossils from stromatolites ranging in
morphology from flat-laminated, or stratiform, to columns. In
addition, vertical transitions between one stromatolite morphology to another are also microfossiliferous and show subtle microbial changes. The stromatolites of the Gunflint are exceedingly variable in their morphology--a characteristic not un-
common to stromatolites.

There are certain forms, however, that are distinctive, less
variable, and have a morphologic "theme." By theme, I refer to
the consistency of morphological features within a bioherm and
recognizable in different outcrops and bioherms I have been able
to recognize four morphologically distinctive Gunflint stromatolites (three columnar and one stratiform; Fig. 3), which were
studied paleomicrobiologically in petrographic thin sections.

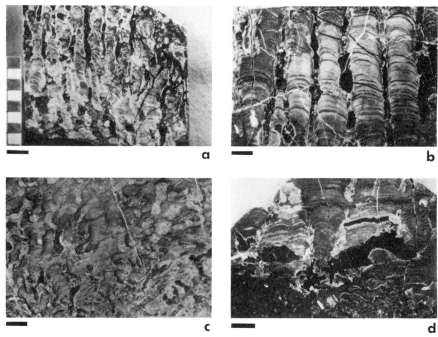

Fig. 3 Gunflint stromatolites. (a) Gruneria biwabikia
(Cloud and Semikhatov, 1969). From Scheiber Beach locality. (b)
?Kussiella from Winston Point. Has Gymnosolen style of branch-
ing. (c) Unnamed columnar stromatolite from Winston Point. May
be referrable to Carelezoon Butin. (d) Stratifera from Winston
Point. Bar equals 1cm.

With the aid of a mechanical point-counting stage, I counted
every microfossil observed in the field of vision under oil
emersion (about 150 μm in diameter) for 0.5 mm intervals, along
both the x and y axes. Fifteen thin sections were examined, and
in excess of 50,000 microorganisms were counted.

The microbiota in the stromatolites consisted predominantly of the filamentous *Gunflintia minuta* Barghoorn and *Huroniospora* sp., all of presumed blue-green algal affinities (Fig. 4).
Other morphotypes are found within the stromatolites but are
rare and statistically insignificant. The three columnar stromatolites contained a microbiota dominated by *G. minuta*, while
stratiform structures contained an assemblage dominated by
Huroniospora sp. (Table 3).

Turbulence did not appear to be a factor in controlling
the morphology of the Gunflint stromatolites. The flat-laminated stromatolites grew in varying degrees of turbulence, as evidenced by the size range of ooids and clasts contained within
laminae. There was no corresponding change in the microbiotic
assemblage reflecting turbulence or the lack of turbulence. The
columnar forms do appear to be associated with turbulence, but
the significance of this is unclear (Table 4).

The biologic changes in the Gunflint stromatolites are subtle, consisting of changes in the relative abundance of microorganisms and not the presence or absence of specific morphotypes
forming different stromatolites. As the microorganism composition of a Gunflint stromatolite changes, the morphology of that
stromatolite also changes. This is clear in the vertial transition from the unnamed stromatolite to *Stratifera*, and then
to ?*Kussiella* through a distance of 4 to 5 cm (Fig. 5). The
community changes from one dominated by *G. minuta* in the unnamed
columnar form to a *Huroniospora* sp. - dominated community in
Stratifera. As the transition continues, *G. minuta* increases
in abundance relative to *Huroniospora*, reaching a proportion of
55% in the well-defined ?*Kussiella* columns (Fig. 6).

The presence of a distinct assemblage within a given stromatolite morphology suggests a biotic control on that stromatolite
morphology (Awramik, 1971b; Licari and Cloud, 1972). No detectable paleoecological changes are associated with these macro-
structural and microbiotic transformations in the Gunflint. However, biologically, one cannot separate ecological changes from
biological changes affecting stromatolite morphology at a given
instant in time. In the Gunflint, presumably minor changes in
the physical environment produced changes in the microbial assemblage and stromatolite morphology.

VI. CONCLUSIONS

The sheer presence of a stromatolite conveys a great deal of

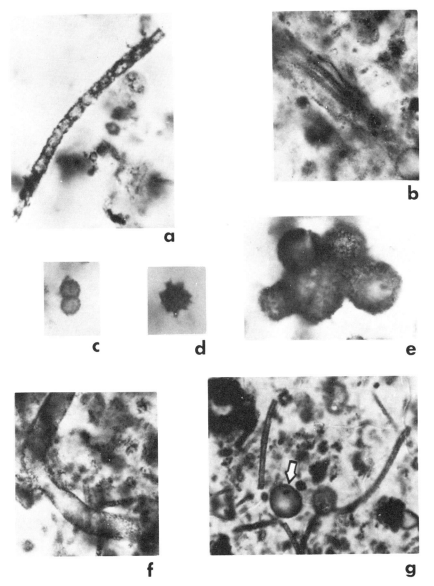

Fig. 4 *Microorganisms found in Gunflint stromatolites.* *(a)*
Gunflintia grandis *Barghoorn.* *(b) Sheath containing two trichomes.*
Trichomes referrable to G. minuta *Barghoorn. (c)* Huroniospora
Dividing. (d) Emicrhystridium barghoorni *Deflandre. (e) Aggregate
of coccoid forms. (f) Empty cyanophycean sheath. (g) G.* minuta
and Huroniospora *sp.* *Note pseudoorganelle within* Huroniospora
(arrow) formed during cyanophycean cell degradation. *Bar equals
10* μm.

TABLE 3

MICROORGANISM CONTENT OF GUNFLINT STROMATOLITES
(numbers and + refer to Braun-Blanquet Scale of Abundance)

	COCCOIDS			FILAMENTS				
	Huroniospora Form 1 (>3μm)	Huroniospora Form 2 (<3μm)	Eomicrhystridium barghoorni Deflandre	Gunflintia minuta Barghoorn	Gunflintia grandis Barghoorn	Animikiea septata Barghoorn	COCCOID CYANOPHYTES	FILAMENTOUS CYANOPHYTES
Gruneria biwabikia	3	1	+	4	+	+	30-34%	66-70%
Kussiella	3	1		4	+		42-46%	54-58%
unnamed stromatolite	3	2		4	1		44-48%	52-56%
Stratifera	4	2		3	+		65-68%	32-35%

Braun-Blanquet Scale

5 = 75%
4 = 50-75%
3 = 25-50%
2 = 5-25%
1 = 5%
+ is rare

125

TABLE 4
Summary of Gunflint Stromatolite Data

Stromatolite	Algae type and content(%)	Range of ooids and clasts maximum apparent Diameter (in mm)	Average maximum diameter of ooids and clasts (in mm)
Gruneria biwabikia	coccoid (30 - 34) filaments (66 - 70)	0.1 - 0.9	0.5
Kussiella	coccoid (42 - 46) filaments (54 - 58)	0.3 - 11	4.0
Stratifera	coccoid (65 - 68) filaments (32 - 35)	0.0 - 1.5	0.8
Unnamed stromato-lite	coccoid (44 - 48) filaments (52 - 56)	0.2 - 2.0	1.2

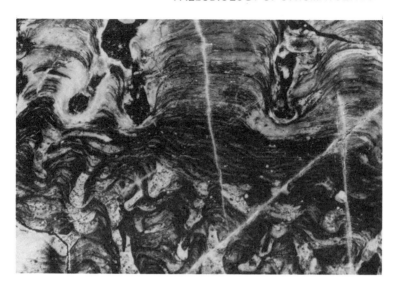

Fig. 5 Transition from unnamed stromatolite to Stratifera
to ?Kussiella. *From Winston Point. Scale is in 0.5-cm intervals.*

paleobiologically useful data. These data, though based on anal-
ogies with Recent algal mats, are consistent with our present
understanding of stromatolites; in particular, that stromatolites
are and were built by communities of microorganisms dominated by
blue-green algae. Stromatolites, early in their history, grew in
environments ranging from intertidal to subtidal. This would require several physiological and behavioral adaptations for the
microorganisms responsible. The increase in morphologic diversity
in columnar stromatolites through the Lower to Upper Riphean
correlates with an increase in taxonomic diversity of microorganisms during the same time span. This may be suggestive of biotic
influence on stromatolite morphology and is strengthened by the
presence of time-restricted assemblages of stromatolite morphologies on a global scale during the Proterozoic.

The microfossiliferous stromatolites of the Gunflint Iron
Formation provide data which aid in interpreting ancient stromatolite morphologies. Subtle microbial differences, in terms of
changing proportions of the different microorganisms available,
can affect morphological changes in stromatolites. These microbial and morphologic changes are probably related to minor, though
at present paleoecologically undetectable, environmental differences. In the Gunflint, a distinctive stromatolite morphology
contains a distinctive microbial assemblage reflecting these ecologically defined microbiological conditions; this may also be
true of other Proterozoic stromatolites. A biological response
or an evolutionary solution to constant environmental conditions

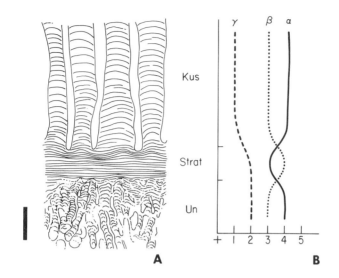

Fig. 6A *Vertical transition from unnamed stromatolite (Un)*
to Stratifera (Strat) to ?Kussiella (Kus).
   Fig. 6B *Changes in mircobiota through transition:* α = *Gun-*
flintia minuta; β = *Huroniospora,* >3 μm; γ = *Huroniospora,* <3
μm. *X axis is Braun - Blanquet (1932) scale of abundance.*

or changing conditions must be considered. Viewing the distribution of stromatolites in time and space during the Proterozoic,
I consider that the evolution of new members and/or the extinction
of existing members in cyanophytic mat-building communities among
taxa, possibly as low as the specific and generic levels, in interaction with the evolving atmosphere and hydrosphere, was responsible for the increase in abundance and diversity and the time-
restriction of certain stromatolite morphologies during the Proterozoic. The end of biostratigraphically useful stromatolite
morphologies during the Cambrian may reflect several factors.
With the rise of metazoans in the late pre-Cambrian the effective
range of stromatolite-building blue-green algae was severly limited. Algal mats could only flourish in areas that excluded or
diminished metazoan activity, such as hypersaline regions and intertidal to supratidal regions. Only those blue-greens that could
adapt to these peripheral environments or evolve this adaptive
ability could survive. This may have resulted in the extinction
of the majority of stromatolite-building taxa   (see Monty, 1973b,
for an alternative explanation.)
   As more data are gathered and our understanding of stromatolite-building processes increases, I hope we may be able to recognize certain ecologically induced biological changes (coupled with
an evolving biosphere, atmosphere, and hydrosphere) and relate
these to such morphologic attributes of stromatolites as style of

branching, convexity of laminae, microstructure, size and shape
of columns, and type and variation of stromatolitic bioherms.
Until that time, the detailed paleomicrobiology of stromatolites
will remain uncertain and open to a certain degree of speculation.

Acknowledgments

*Financial support for field work associated with the ideas
expressed in this paper was supplied by the Department of Geological Sciences (Harvard University), the Committee on Evolutionary
Biology (Harvard), NSF Grant GA 37140 to E. S. Barghoorn (Harvard), two Grants-in-Aid of Research from Sigma Xi, and the Academic Senate of UCSB. Discussions with E. S. Barghoorn, P. E.
Cloud, S. Golubic, L. Margulis, and M. R. Walter were most helpful. D. Crouch, UCSB, drafted the figures.*

REFERENCES

Awramik, S. M. 1971a, *Science*, *174*:825.

Awramik, S. M. 1971b,*Geol. Soc. Amer. Abstr. Prog.*, *3*(7):496.

Awramik, S. M. In M. R. Walter (ed.), *Stromatolites*, chap. 4.4.
Elsevier, Amsterdam, 311.

Awaramik, S. M., and Barghoorn, E. S. 1975. *Geol. Soc. Amer.
Abstr. Prog.*, *7*(3):291.

Awramik, S. M., Margulis, L., and Barghoorn, E. S. 1976, In M. R.
Walter (ed.), *Stromatolites*, chap. 4.2. Elsevier, Amsterdam,
149.

Barghoorn, E. S., and Tyler, S. A. 1965. *Science. 147*:563.

Bertrand-Sarfati, J. 1972. *Cent. Nat. Rech. Ser. Geol.*, *14*:242.

Braun-Blanquet, J. 1932. *Plant Sociology.* McGraw-Hill, New York.

Butin, R. V. 1966. In *Ostatki organizmov i problematika proter-
ozoiskikh obrazovanii Karelii*, p. 34. Petrozovodsk.

Button, A. 1971. *Trans. Geol. Soc. S. Afr.*, *74*:201.

Cloud, P. E. 1965. *Science. 148*:27.

Cloud, P. E. 1976b.*Major Features of Crustal Evolution, Trans.
Geol. Soc. S. Afr.*, *71:1*.

Cloud, P. E., Licari, G. R., Wright, L. A., and Troxel, B. W
1969. *Proc. Natl. Acad. Sci. U.S.A*, *62*:623.

Cloud, P. E., and Semikhatov, M. E. 1969. *Amer. J. Sci.*, *267*:1017.

Diver, W. L. 1974. *Nature (London) 247*:361.

Donaldson, J. A., and Delaney, G. 1975. *Canad. J. Earth Sci.*,
12:371.

Gebelein, C. D. 1974. *Amer. J. Sci.*, *274*:575.

Ginsburg, R. N., Isham, L. B., Bein, S. J., and Kuperberg, J.
1954. Unpublished report No. 54-21, Marine Laborotory, University of Miami, Corol Gables, Florida.

Golubic, S. 1973. In N. G. Carr, and B. A. Whitton (eds.), *The
Biology of Blue-Green Algae*, pp. 434 - 472. University of

California Press, Berkeley, Calif.
Golubic, S., and Hofmann, H. J. J.   *leontol., in press.*
Hofmann, H. J. 1969a. *Geol. Surv. Canada. Paper 68 - 69:1.*
Hofmann, H. J. 1969b. *Geol. Surv. Canada. Paper 69 - 39:1.*
Hofmann, H. J. 1973. *Earth-Sci. Rev., 9(4):339.*
Hofmann, H. J. 1974. *Nature (London), 249:87.*
Hofmann, H. J., and Jackson, G. D. 1969. *Canada J. Earth Sci.,*
    6:1137.
Keller, B. M., Kazakov, G. A., Krylov, I. N., Nuzhnov, S. V.,
    and Semikhatov, M. A. 1960. *Izv. Akad. Nauk SSSR Ser.Geol.,*
    12:26.
Kirichenko, G. I. 1964. *Sov. Geol., 6.*
Knoll, A. H., Barghoorn, E. S., and Golubic, S. 1975. *Proc. Natl.*
    Acad. Sci. U.S.A., 72:52.
Korde, K. B. 1953. *Dokl. Akad. Nauk SSSR, 89:1091.*
Korolyuk, I. K. 1960. *Trudy Inst. Geol. Razrab. Goryuch. Iskop.*
    Akad. Nauk SSSR, 1, 112.
Korolyuk, I. K. 1963. In B. M. Keller (ed.), *Stratigrafiya SSSR,*
    Verkhnii dokembrii, pp. 479 - 498. Gosgeol-tekhizdat, Moscow.
Krinksy, N. I. 1966. In T. W. Goodwin (ed.), *Biochemistry of*
    Chloroplasts, p. 423. Academic Press, New York.
Krylov, I. N. 1963. *Akad. Nauk SSSR, Trudy Inst. Geol. 69:133.*
Krylov, I. N. 1967. *Akad. Nauk SSSR, Trudy Inst. Geol.*
    171:88.
Leff, J., and Krinsky, N.I.T. 1967. *Science, 158:1322.*
Licari, G. R. *J. Paleontol., in press.*
Licari, G. R., and Cloud, P. E. 1972. *Proc. Natl. Acad. Sci. USA,*
    69:2500.
Licari, G. R., Cloud, P. E., and Smith W. D. 1969. *Proc. Natl.*
    Acad. Sci. USA, 62:56.
Logan, B. W., Hoffman, P., and Gebelein, C. D. 1974. *Amer. Assoc.*
    Petrol. Geol., 22:140.
MacGregor, I. M., Truswell, J. F., and Eriksson, K. A. 1974.
    Nature (London), 247:538.
McKirdy, D. M. 1974. *Precambrian Res., 1:75.*
Monty, C. L. V. 1967. *Ann. Soc. Geol. Belgique, 90:55.*
Monty, C. L. V. 1973a. *C. R. Acad. Sci., 276(D):3285.*
Monty, C. L. V. 1973b. *Ann. Soc. Geol. Belgique, 96:585.*
Muir, M. D. 1974. *Origins Life, 5:105.*
Muir, M. D. *Alcheringa, in press.*
Nagy, L. A. 1974. *Science, 183:514.*
Preiss, W. V. 1972. *Trans. Roy. Soc. S. Austral.,96:67.*
Preiss, W. V. 1973. *Trans. Roy. Soc. S. Austral.,97:91.*
Raaben, M. E. 1969. *Akad. Nauk SSSR, Trudy Inst. Geol., 203:100.*
Raaben, M. E., and Komar, V. A. 1964. *Izv. Akad. Nauk. SSSR, Ser*
    Geol., 6:109.
Schopf, J. W. 1968. *J. Paleontol., 42:651.*

Schopf, J. W. 1975a. *Ann. Rev. Earth Planetary Sci.*, *3*:213.

Schopf, J. W. 1975b. In *Chemical Evolution of the Precambrian*, pp. 30 - 33. Laboratory of Chemical Evolution, University of Maryland, Abstracts.

Schopf, J. W., and Barghoorn, E. S. 1969. *J. Paleontol.*, *43* ¦11.

Schopf, J. W., and Blacic, J. M. 1971. *J. Paleontol.* *45*:925.

Schopf, J. W., and Fairchild, T. R. 1973. *Nature (London)*, *242*: 537.

Schopf, J. W., Oehler, D. Z., Horodyski, R. J., and Kvenvolden, K. A. 1971. *J. Paleontol.*, *45*:477.

Semikhatov, M. A. 1962. *Akad. Nauk SSSR, Trudy Inst. Geol. 68*:242.

Semikhatov, M. A. 1974. *Akad. Nauk SSSR, Trudy Inst. Geol. 256*:302.

Semikhatov, M. A., and Komar, V. A. 1965. *Dokl. Adad. Nauk SSSR, 165*:1383.

Semikhatov, M. A., and Komar, V. A., and Nuzhnov, S. V. 1963. In *Teginalnaya stratigrafiya SSSR*, pp. 32 - 44. Gosgeoltekhizclat, Moscow.

Serebryakov, S. N. 1975. *Akad. Nauk. SSSR, Trudy Geol. Inst. 200*:1

Truswell, J. F., and Eriksson, K. A. 1972. *Trans. Geol. Soc. S. Afr.*, *75*:85.

Valdiya, K. S. 1969. *J. Geol. Soc. India*, *10*:1.

Vologdin, A. G. 1955. *Priroda*, *9*:39.

Vologdin, A. G. 1962. *Drevneyshie Vodorosli SSSR Izd.-vo Akad. Nauk SSSR Moscow*, 656 pp.

Walter, M. R. 1972. *Palaeontol. Assoc. Spec. Pap.*, *11*:190.

Walter, M. R., and Preiss, W. V. 1972. *Int. Geol. Congr.*, *Sect. 1, Precambrian Geology*, pp. 85 - 93.

14

NATURAL MECHANISMS OF PROTECTION OF A BLUE-GREEN ALGA AGAINST ULTRAVIOLET LIGHT

MITCHELL RAMBLER and LYNN MARGULIS
Boston University
ELSO S. BARGHOORN
Harvard University

Lyngbya *is a very common genus of blue-green algae, often the surface component of algal mat communities. Our studies show that the uv tolerance of* Lyngbya sp. *can be extended from the lethal value (1.46 x 10^3 ergs/mm^2, or 10 min at 2537°A) to at least 24 hr (2.0 x 10^5 erg/mm^2) of continuous uv irradiation. Protection is conferred (in the absence of conditions permitting photoreactivation) by the addition to the media during irradiation of sodium nitrate and nitrite salts (which have strong uv absorption properties). Several days of protection against potentially lethal uv can be achieved simply by adjusting inoculum weight (from 1.1 to 9.1 mg), i.e., taking advantage of the colonial matting habit of the microorganism. It is possible that nitrate- and nitrite-mediated ultraviolet light protection preadapted Precambrian microorganisms for the anaerobic respiration of these substances, thus leading to the origin of nitrite and nitrate reduction. (The end products of these metabolic pathways are nitrogen and nitrous oxide which presumably would have been eliminated into the atmosphere.)*

I. INTRODUCTION

In preliminary studies of *Lyngbya* sheaths taken from the surface of algal mat communities at Sippewissett salt marsh (Cape Cod, Massachusetts) a striking uv absorption feature was noted in aqueous solutions of sheath material. It had a maximum at approximately 250 mμ. This feature was repeatedly seen both in preparations made from natural populations of algae (collected from near the mouth of the Charles River and the

133

Fens, Boston, Massachusetts) and in laboratory-grown specimens. Further investigations revealed that this absorption was due primarily to the presence of salts in the various aqueous media surrounding the algae. These salts presumably entered the algal sheaths under study. These observations suggested to us that common environmental substances abundantly present in algal mat communities might have conferred protection against solar uv at times in the past in which shorter wavelength radiation is thought to have reached the Earth's surface.

The concept that the colonial habit characteristic of sheathed algal communities is, in part, protective against light has often been suggested (e.g., Monty, 1967; Korde, 1972). To our knowledge, however, this is the first report of experiments designed to estimate the protective value of nitrate and nitrate salts and inoculum size in blue-green algae exposed to ultraviolet light.

Stromatolites are defined as layered organosedimentary structures built by microscopic algae and bacteria (Walter, 1972); commonly they are lithified remains primarily of blue-green algal mat communities (Monty, 1967). These structures are distributed worldwide; in time they range from the early Precambrian to the present (MacGregor, 1941; Bond et al., 1973). In recent years, due to research on stromatolites and associated microbiotas, there has been a remarkable growth in our understanding of microbial contributions to the Precambrian environment. Since the pioneering work of Barghoorn and Tyler (1954) on microscopic thin sections through algal cherts, the recognition that the Precambrian was primarily the age of the procaryotic microbes has developed apace (Schopf and Barghoorn, 1967; Schopf et al., 1971). In the earliest sediments of southern Africa there is ample evidence that microbial communities flourished >3000 million years ago (Engel et al., 1968; Van Niekerk and Burger, 1969) and that substantial diversity had arisen in blue-green algae (≡cyanobacteria) by 2200 million years ago (Nagy, 1974). Although the community of microbes responsible for the construction of the algal mat precursors of stromatolites may range from photosynthetic bacteria (e.g., Chloroflexis; Brock, 1969) to a wide variety of blue-green algal species, both the matting habit and photosynthesis had probably evolved by 2700 million years ago, e.g., by the time the Bulawayan stromatolites of southern Africa had formed. According to Cloud (1974), this was prior to the formation of the oxygenic atmosphere and thus before the development of the uv-absorbing stratospheric ozone layer.

The results we report here are of studies designed to measure the tolerance and protection of pure cultures of Lyngbya sp. irradiated with uv light. Observations of living algal mat communities have shown a biostratigraphic zonation correlated with vertical differentiation of energy levels and external environmental parameters (Golubic, 1973). *Lyngbya* was used because it occupies the uppermost portion of the local mats, possibly providing a protective layer for microbial populations in the lower portions of the community. Our results have suggested to us that the matting habit and the retention of certain oxidized nitrogen salts (produced as a result of blue-green algal oxygen elimination) may have originally been selected for because of their protective value against the potentially lethal ultraviolet light. It is very likely that solar uv reached the Earth's surface during the Archean, prior to the formation of the ozone layer.

II. MATERIALS AND METHODS

The ultraviolet absorptions of aqueous solutions of sodium nitrate and sodium nitrite were calibrated by a Perkin-Elmer 202 recording drum UV-Vis spectrophotometer. A 1.0-cm light path was used. All irradiations were performed in 3.0 ml of liquid media at a maximum depth of 1.0 cm. The extent to which relatively low concentrations of these salts can entirely filter out the ultraviolet from about 200 to >250 nm can be seen in Fig. 1.

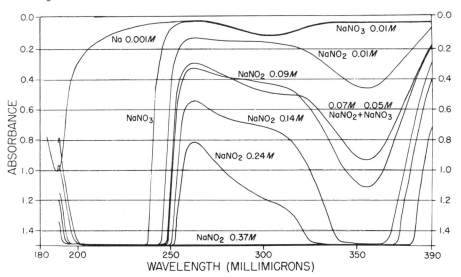

Figure 1   Ultraviolet light absorption spectra of aqueous solutions of sodium nitrate and nitrite.

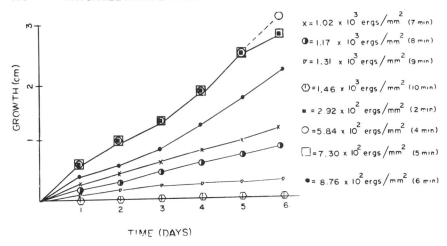

Figure 2   Determination of lethal uv doses of unprotected
Lyngbya *sp.*; Growth after irradiation.

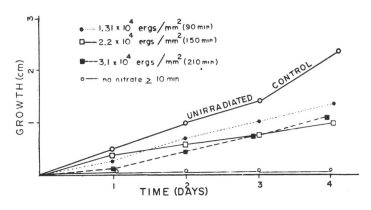

Fig.   3   Growth after potentially lethal uv irradiation
of sodium nitrate protected Lyngbya.

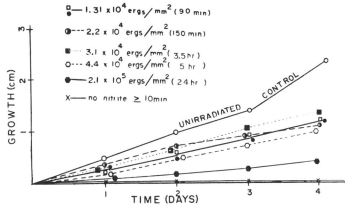

Fig.   4   Growth after potentially lethal uv irradiation
of sodium nitrite protected Lyngbya.

Since most cell death by uv is due to the absorption of light by nucleic acids and proteins (λ_{max} = 260 - 280 nm), the potential for protection of cells by high concentrations of these simple nitrogenous salts was suggested simply by the features of the absorption spectrum.

The filamentous blue-green alga, *Lyngbya* sp., was supplied by the Indiana University Culture Collection, no. 621. Stock cultures were grown under sterile conditions in aqueous BG-11 medium in 250-ml Ehrlenmeyer flasks. The medium BG-11 (Stanier *et al.*, 1971) was modified from G-11 (Hughes, 1958). BG-11 consists of: $NaNO_3$, 1.5 g/liter; K_2HPO_4, 0.04 g/liter; $MgSO_4$ · $7H_2O$,0.075 g/liter; $CaCl_2$· $2H_2O$,0.036 g/liter; Na_2CO_3, 0.02 g/liter; citric acid · $1H_2O$,0.0065 g/liter; ferric ammonium citrate (brown), 0.006 g/liter; EDTA (disodium magnesium salt), 0.001 g/liter; trace metals: H_3BO_3 2.86 mg/liter; $MnCl_2$ · $4H_2O$, 1.81 mg/liter; $ZnSO_4$ · $7H_2O$,0.222 mg/liter; Na_2MoO_4 · $2H_2O$, 0.39 mg/liter; $CuSO_4$ · $5H_2O$,0.079 mg/liter; $Co(NO_3)_2$ · $6H_2O$,0.049 mg/liter (pH is adjusted to 7.8). All irradiated samples were inoculated with 1.1 or 9.1 mg of inocula placed on solid BG-11 mineral agar in plastic (35 x 10 mm) Petri dishes.

The *Lyngbya* inocula from sterile pure cultures were irradiated using a G30T8 germicidal mercury vapor lamp fitted in a Lab Con Co. radiation hood. A Rohm and Haas yellow interference filter (No. 2422) was placed directly on top of the cultures to prevent photoreactivation (Van Baalen, 1968; Van Baalen and O'Donnell, 1972). Because microorganisms placed in white light immediately after uv radiation are known to photorepair, throughout all of these experiments photoreactivation in the algae was prevented. Even during postirradiation growth, the photoreactivating white light was filtered out. After irradiation, all cells were incubated continuously in filtered white light for days while growth was monitored. Thus, our results consistently record minimum protection conferred by the agents tested. Under natural conditions, of course, photoreactivating white light would be expected to be mixed in with the potentially lethal ultraviolet light. Visible light measurements were made with a Western Master No. 560 light meter. The controls were similar-sized inocula placed on the same solid media without irradiation. After irradiation, the lids were placed on the dishes and cultures were placed in a northern exposure window under Sylvania "Gro-Lux" lamps (F40-GRO-WS) supplying 125 foot-candles per square foot and grown at approximately 25°C for a period of about one week.

Sodium nitrate protection of the algae was studied by placing a standardized 1.1 mg of inoculum in the small Petri dishes containing 3.5 ml of test solution. BG-11 medium minus all sodium compounds was used for the control. Normal BG-11 plus nitrate or nitrite at appropriate concentrations was used for the experimentals. Unless otherwise stated, the $NaNO_3$ concentrations used for the protective media were 0.37 M. Growth of irradiated and control cultures in normal BG-11 cultures was monitored for about 1 week after uv treatment by measurements of radial extension of the trichomes.

The effect of increasing the inoculum size from 1.1 to 9.1 mg has also proven to be dramatic. Although a 1.1-mg inoculum is killed with 10 min of irradiation, an unprotected cellular clump or mat of 9.1 mg can withstand irradiation at least 48 hr (4.2 x 10^5 erg/mm^2, Fig. 5).

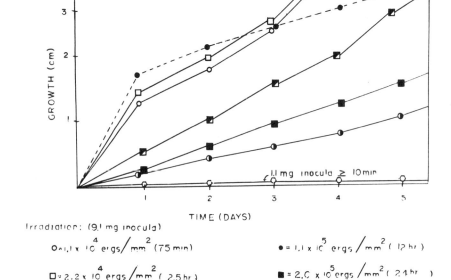

Irradiation: (9.1 mg inocula)

O = 1.1 x 10^4 ergs/mm^2 (75 min)

$\square$ = 2.2 x 10^4 ergs/mm^2 (2.5 hr)

$\blacksquare$ = 5.7 x 10^4 ergs/mm^2 (6.5 hr)

● = 1.1 x 10^5 ergs/mm^2 (12 hr)

■ = 2.0 x 10^5 ergs/mm^2 (24 hr)

◑ = 4.1 x 10^5 ergs/mm^2 (48 hr)

Fig. 5 Growth after potentially lethal uv irradiation of 9.1 mg inocula of Lyngbya.

Moreover, combining both salts with large inoculum size increases the tolerance of *Lyngbya* sp. to at least 3 days of continual irradiation (or 6.3 x 10^5 erg/mm^2). Our observations suggest that irradiated inocula form a matted layer of surface cell debris, that is, a crust which protects the underlying algae from further loss of viability.

Although after irradiation in protective media, growth of filaments is slowed somewhat (Figs. 3 - 5), irradiated inocula are completely viable. The patterns of postirradiation growth are variable; they do not show a simple retardation rate as a function of irradiation probably, in part, because complex repair systems may be activated. In spite of variable postirradiation growth, survival was as depicted in Figs. 3 - 5 for at least three trials per point. The simple linear trichome growth

measurements here were only made for the purpose of verifying
colony viability; for detailed understanding of the mechanism
of radiation protection more quantitative studies would have
to be undertaken. Taken together, however, our results show unequivocally that even pure cultures of surface blue-green algae
characteristic of modern algal mats can withstand more than 2
days worth of potentially lethal continuous ultraviolet irradiation. Furthermore, even in the absence of active photoreactivation systems they can protect themselves by relatively simple
and available natural means: the uptake of nutrient salts from
the medium and the development of the colonial matting habit.
Thus, it is likely that even in the Precambrian, in the absence
of an ozone layer, photosynthetic organisms would not necessarily be limited to a 1 m depth of ocean or areas protected by
shade. High uv fluxes would probably not per se prevent the
development of species diversity and the occupation of new niches.
Selection pressures would probably lead to the maintenance of
certain concentrations of the sodium nitrates and nitrites and
the development of matted growth patterns. Most blue-green algae are obligatory photoautotrophs; yet, it is likely that they
functioned in shallow aqueous media exposed to ultraviolet radiation for the entire solar day during the Precambrian. Other repair systems during alternating periods of darkness must also
have helped maintain viability (Zhevner and Shestakov, 1972).
Even in shallow isolated pools subject to evaporation or tidal
influences some blue-greens could survive, since 24-hr protection
has been shown by 9.1-mg clumps. Our data and many other studies make it seem unlikely that the rise of "higher organisms"
was directly limited in the Precambrian by high uv fluxes and low
oxygen tensions as postulated by Fischer (1965; 1972) and Berkner and Marshall (1964, 1966). A reevaluation of the role of
uv and oxygen in the origin of higher organisms is in progress
(Rambler et al., 1976).
     The widespread ability for reduction of nitrate among the
bacteria is well known. The utilization of nitrate as electron
acceptors may have evolved in microbes subsequent to the utilization of these substances for uv protection. The metabolic advantages and possible primitive sources of nitrate have been discussed in detail (Hall, 1971). The role of nitrates in biological oxidations and enzymatic mechanisms involved have also been
recently reviewed (Egami, 1973, Broda, 1975; Payne, 1973. Our
work seems consistent with the contention of Egami (1976) that
the evolution of nitrate reduction pathways preceeded in microbes
the origin of aerobic respiration. It is reasonable to postulate
that the evolution of widespread oxygen-eliminating metabolism
occurred in cyanobacteria ($\equiv$blue-green algae; see Cohen et al.,
1975, for a report on facultative oxygen-producing microbial
photosynthesizers). Furthermore, the increasing elimination of

molecular oxygen probably provided the oxidant leading to local production of soil nitrates and nitrites. These compounds may have originally been concentrated for uv protection in microbes; such organisms may subsequently have evolved anaerobic respiratory (nitrate reduction) pathways. It is interesting to note that the metabolic end products of these respiratory pathways are molecular nitrogen and nitrous oxide. Thus, the potential for altering the gaseous composition of the Earth's atmosphere probably evolved in the early Precambrian.

Acknowledgments

    Support for this research came from NASA NGR-004-025 (to L. Margulis), NASA NGL-22-007-069, and NSF GA 13821 (to E. S. Barghoorn).

REFERENCES

Barghoorn, E. S., and Tyler, S. A. 1965. *Science, 149*:563.
Berkner, L. V., and Marshall, L. C. 1964. *Discuss. Faraday Soc. 39*:122.
Berkner, L. V., and Marshall, L. C. 1966. *J. Atmos. Sci., 23*:133.
Bond, G., Wilson, J. F., and Winnall, N. J. 1973. *Nature (London), 244*:275.
Brock, T. D. 1969. *Phycologia, 8*:201.
Broda, E. 1975, *The Evolution of Booenergetic Process,* Pergamon Press, Oxford.
Cloud, P. 1974. *Amer. Sci., 62*:54.
Cohen, Y., Palen, E., and Shilo, M. 1975. *J. Bacteriol.,* p. 855.
Egami, F. 1976. *Origins Life,* Vol. 7: 71 - 72.
Egami, F. 1973. *Z. Allg. Mikrobiol. 13*:177.
Engel, A. E., Nagy, B., Nagy, L. A., Engel, C. G., Kremp, G., and Drew, C. M. 1968. *Science, 161*:1005.
Fischer, C. 1965. *Proc. Natl. Acad. Sci. USA, 53*:1205.
Fischer, C. 1972. *Main Comments in Modern Thought, 28*:159.
Golubic, S. 1973. In Carr and Whitton (eds.), *The Biology of Blue-Green Algae.* University of California Press, Berkeley.
Hall, J. 1971. *J. Theor. Biol., 30*:429.
Korde, K. B. 1972. In T. V. Desikachary (ed.), *Taxonomy and Biology of Blue-Green Algae.* Indian Council of Agricultural Research, New Delhi.
Macgregor, A. M. 1941. *Trans. Geol. Soc. S. Afr. 43*:9.
Monty, C. L. V. 1967. *Ann Soc. Geol. Belg., 9*:55.
Monty, C. L. V. 1971. *Ann. Soc. Geol. Belg., 94*:265.
Nagy, L. A. 1974. *Science, 183*:514.
Payne, W. J. 1973. *Bact. Review. 37*:409 - 450.

Rambler, M., Margulis, L., and Walker, J. C. G. 1976, *Nature* (in press).

Rupert, C. S. 1974. *Photochem. Photobiol.* *2*:205.

Schopf, J. W. 1968. *J. Paleontol.* *42*:651.

Schopf, J. W., and Barghoorn, E. S. 1967. *Science,* *156*:508.

Schopf, J. W., Oehler, D. Z., Horodyski, R. J., and Kvenvolden, K. A. 1971. *J. Paleontol.,* *45*:477.

Van Baalen, C. 1968. *Plant Physiol.* *43*:1689.

Van Baalen, C., and O'Donnell, R. 1972. *Photochem. Photobiol.,* *15*:269.

Van Niekerk, E. B., and Burger, A. J. 1969. *Trans. Geol. Soc. S. Afr.* *72*:9.

Zhevner, V. D., and Shestakov, S. 1972. *Sanv. Arch. Mikrobiol.* *86*:349.

15

KAKABEKIA, A REVIEW OF ITS PHYSIOLOGICAL AND ENVIRONMENTAL FEATURES AND THEIR RELATION TO ITS POSSIBLE ANCIENT AFFINITIES

BARBARA Z. SIEGEL
University of Hawaii at Manoa

Kakabekia barghoorniana *was named after the Precambrian* Gunflint microfossil, *Kakabekia umbellata*, *first described in 1965 by Barghoorn and Tyler* (Science, 147:563). *K. barghoorniana was initially recognized on a nutrient agar plate which had been inoculated with a soil sample from Wales and maintained in an NH$_3$ atmosphere at 760 mm Hg. Subsequently, it has been cultivated from soil inocula in glucose - 15 M aqueous NH$_3$ media. Its growth rate is very low and it has yet to be grown in pure culture. Subsequent finds of the living form include sites in Alaska, Iceland, Japan, and Hawaii above 1300 m. This chapter will review cytological, physiological, and environmental parameters of living* Kakabekia *with respect to its possible ancient affinities as well as to its status as a highly unusual microorganism, whatever its affinities prove to be.*

The genus *Kakabekia* consists of two recognized species, *K. umballata* (Fig. 1) and *K. barghoorniana* (Fig. 2). The former is known only as a microfossil from cherts of the Gunflint Iron Formation in southern Ontario. These cherts, which contain other microfossils, as well, have been described in detail and assigned to the Middle Precambrian, ca. 2 x 10^9 years of age (Barghoorn and Tyler, 1965). *K. barghoorniana* is a rare living member of the genus and was first discovered in soils from Harlech, Wales in 1964 (Siegel and Giumarro, 1966). Since its discovery, *Kakabekia* has been grown routinely in 15 *M* aqueous ammonia and was considered to be an obligate ammonophile. It has, however, never been cultured in pure form independent of the initial soil inoculum. Furthermore, growth is slow and yields

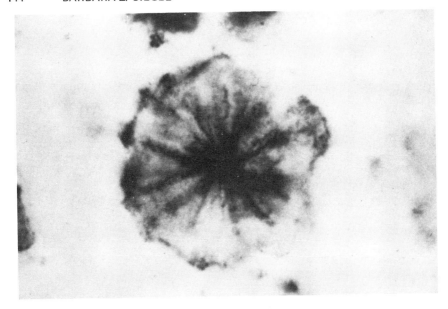

Fig. 1   *Fossil* Kakabekia umbellata, *thin section of the*
Gunflint chert deposit.   The organism is approximately 10 µm
in diameter.   Photographed at 795X by E. Barghoorn.

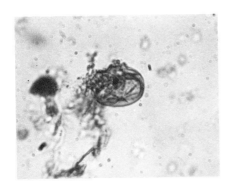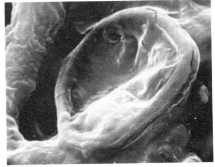

Fig. 2   *Left photomicrograph of an unstained specimen*
photographed at 567X.   The right is a scanning electron micro-
scope photograph at 1254X.   Both pictures are of Kakabekia
barghoorniana *and have cell diameters of about 10 µm.*

of biomass are low, such that it has never been obtained in
sufficient numbers for comprehensive biochemical analyses.   It
has been possible, nevertheless, to establish a number of significant characteristics which include details of cytomorpholo-
gy, the presence or absence of heme-containing enzymes, growth

requirements (e.g., ammonia, oxygen, and temperature), and its probably procaryotic nature.

After 10 years of study, in many respects *Kakabekia* is as perplexing as it was in 1965. The assignment of the organism sirst cultured from soil collected in Wales to the genus *Kakabekia* seems fully reasonable on the basis of optically resolvable structural detail, its procaryotic organization, and the population variability in morphology and heme enzyme content. Recently, several workers have reported observing flat star-shaped bacteria with tapering prothecae (Hirsch, 1974; Nikitin, 1966, 1973; Vasiliva, 1970). These "stars" are loosely grouped taxonomically with the gram-negative bacteria which bud, are grown from sources rich in humic matter such as soil suspensions or creek water, and, except for one Russian report (Vasiliva, 1970), have not yet been grown in pure culture. Hirsch (1974) has said that although *Kakabekia* is nearly twice as large, "a certain morphological resemblance exists between these star-shaped bacteria...and *Kakabekia*." Furthermore, it is reported that these budding bacteria inhabit many extreme locations and often show great resistance to stress conditions. In fact, all of the star-shaped bacteria have been collected from relatively cool locations. Like *Kakabekia,* they too may have a general cryophilic growth requirement.

Up until 1969, only samples from the original site at Harlech, Wales yielded *Kakabekia;* unsuccessful surveys were made of many North American and West European soil samples. In that year, *K. barghoorniana* was again obtained--two collections from Alaska produced this organism, one from the flanks of the Mendenhall Glacier, Juneau, and the other from the Point Barrow area well north of the Arctic circle. Thus, the location of possible sampling sites suggested a distribution pattern restricted to higher latitudes northward of 45°N. The possibility that *Kakabekia* was a cryophile, or at least flourished at a low mean temperature, was supported when living *Kakabekia* was found in Iceland in 1970. With this hypothesis as a guide, *Kakabekia* has been sought and found at higher elevations of north temperate and tropical latitudes. The distribution of *Kakabekia* in Iceland was also reexamined in 1972 and 1975.

It is with the results of the more recent studies as they reflect upon the ecological and physicochemical limits of *barghoorniana* that this chapter is principally concerned. Even if the probability of a close and direct affinity between the microfossil and the living form is small, it is more than sufficient to warrant the continued study of *Kakabekia*.

I.   MATERIAL AND METHODS

A.   *Collection Sites*

Soil samples have been obtained from a large number of sites, including the continental United States, north and southeastern Alaska, Hawaii, Mexico, Panama, the Guayanas, Iceland, Wales, France, Belgium, West Germany, Japan, Malaysia, Taiwan, Nepal, and selected areas of South America. These sites ranged from the equator to 70°N latitude and from sea level to nearly 5000 m.

Samples were collected in the field and were generally sealed in plastic. If possible, they were initially air dried and cooled or refrigerated until use. Sites were sampled randomly at the beginning of the studies, but subsequently greater success in detecting *Kakabekia* resulted when more humic soil samples in and around other vegetation were selected.

B. *Culture Conditions*

Routine cultures consisted of 0.5 g of soil in 10 ml of 15 *M* aqueous ammonia containing 0.5% glucose. Inoculated media were incubated at 23°C in filled, tightly stoppered vials, excluding most of the free air space. Experimental variations in temperature, alkali, etc., will be described as appropriate later in this chapter.

C. *Special Procedures*

Mercury analyses were carried out by nitric acid/perchloric acid digestion and flameless atomic absorption as previously described (Siegel *et al.*, 1973). This procedure yields "available" mercury in the organic soil fraction, but does not release mercury immobilized in the silicate matrix of rock or lava residues.

II. RESULTS AND DISCUSSION

A. *Biogiographic Considerations*

An examination of the distribution pattern of *Kakabekia* suggested that an obvious limiting latitudinal factor (Table 1) might well be temperature related. At lower elevations, *Kakabekia* has not been found south of approximately 45°N and hence appears restricted to regions of relatively cool summers. This limitation does not apply to higher elevation, but the presence of *Kakabekia* at sites 2000 m above sea level in tropical as well as lower temperate latitudes is fully consistent with the notion of cryotolerance. Air temperature decreases approximately 1°C for each 140 - 195 m of altitude (higher lapse rates at higher latitudes).

The high latitude "rule" seems to fit recent samples quite

TABLE 1
The Effects of Latitude and Altitude on the Occurrence of
Kakabekia

Location	Latitude (°N)	Altitude (M)
Alaska	60 - 70	--
Iceland	63 - 68	--
Mount Showa Shinzan, Japan	45	--
Sapporo, Japan	43	--
Mount Fuji, Japan	35	2300
Mauna Kea, Hawaii	19	2400 - 4600
Mount Haleakala, Maui	21	2800 - 3000
Himalayan foothills	28	3500
Near Mexico City, Mexico	19	2000
Bogota, Columbia	5	2880
Near Quito, Equador	0	2500

well (Table 2), although its narrow zone on Mount Fuji indicates
that there are other factors operating as well.

TABLE 2
Local Occurrence of Kakabekia at Sample Sites in Japan[a]

Location	Presence of Kakabekia
Hokkaido (latitude approximately 45°N)	
Lake Toya	--
Mount Showa Shizan	+
Sapporo	+
Samani	-
Honshu (latitude approximately 35° N)	
Akiyoshido (cave)	-
Lake Hakone area	-
Mount Fuji	
2300 m	+
3000 m	-
3300 m	-
3750 m	-

[a]Many of these samples were collected by Mr. and Mrs.
James Barres.

Although the vertical pattern on Mauna Kea was also consistent when samples were incubated at -7 to 25°C, incubation at 30°C revealed the presence of an anomalous thermal strain only at the lowest sample station (Fig. 3).

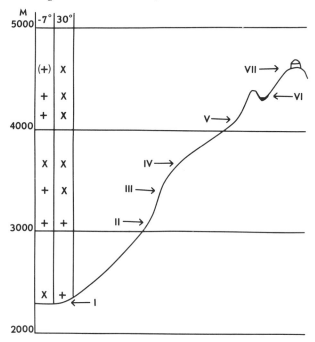

Fig. 3   Vertical distribution of Kakabekia barghoorniana *on Mauna Kea.   The two thermal strains are denoted as -7°C and 30°C.   Presence of* Kakabekia *is denoted by + and its absence by X.   The symbol (+) denotes the presence, but rarity (<4/cm²). Sample stations are located by Roman numerals, elevations in meters.*

This aberrant strain has not been further studied.   It is evident, however, that the more typical form of *Kakabekia* grown at -7°C (and 25°C) is quite sensitive to moderate warmth.

The operation of factors other than temperature in the determination of the vertical distribution pattern is evident on Mount Haleakala (Table 3).   *Kakabekia* has been found only above 2000 m, but it appears that specific edaphic or other habitat factors affect its distribution.

Additional field data confirming sensitivity to heat in *Kakabekia* were obtained from sediment samples flanking the runoff channels from the geyser Stikkur at Geysir, Iceland (Table 4). The area is rich in alkaline thermal waters, but relatively low air and ground temperatures lead to rapid cooling and steep

TABLE 3
Local Occurrence of Kakabekia *along the Slope of Mount Halea-kala (Maui)*

Elevation (m)	Habitat	Presence of Kakabekia
0	Cool cave, moss hummocks	--
700	Eucalyptus forest	--
2000	Rocky scrub land	--
2000	Crevices in outcrops	--
2800	Dry moss hummock on black sand	--
2800	Around roots of *Plantago lanceolata*	+
2800	Shade, moist crevice, lava outcrop	+
3000	Overlook at crater rim, open, black sand	-
3000	Base of mixed stand of grasses and *Plantago*	+

TABLE 4
Occurrence of Kakabekia *in Runoff Streams at Geysir, Iceland*[a]

Approximate distance from source (m)	Water temperature (°C)	Kakabekia in sediment edging stream bed
50	48 – 50	-
100	35 – 38	-
200	22 – 24	+

[a]Sampling of June 24, 1972. Air temperature about 7°C, pH of runoff water was approximately 10.

temperature gradients. *Kakabekia* was in evidence only at locations where the water temperature had fallen below 30°C.

B. *Regional Variations in Morphology*

It was previously reported (Siegel and Siegel, 1970) that the distribution of morphological characters in populations of *Kakabekia* varied with the collection site. Continued attention has been given to this feature (Table 5).

The most conspicuous distinction is the absence of deeply incised mantles in any group, except for the fossil specimens

TABLE 5
Morphological Patterns in Populations of Kakabekia *from Various Sources*

Source	Mantle (%) Incised	Lobed	Entire	Rays (%) 1 - 3	>4	Stipe present (%)
Fossil						
(Ontario)	57	20	23	20	80	50
Pacific Group[a]						
Alaska	0	67	33	20	80	21
Japan	0	59	41	14	86	17
Mauna Kea	0	55	45	11	89	22
Haleakala	0	40	60	12	88	12
Atlantic Group[b]						
Iceland	0	27	73	10	90	25
Wales	0	24	76	8	92	36

[a]Alaska from the Mendenhall Glacier, Juneau; Japan from 2300-m level, Mount Fuji; Mauna Kea from 3200-m level, Hawaii; Haleakala from 2800 m Maui.

[b]Iceland from the Reykjanes Peninsula north of Reykjavik; Wales from the Harlech Castle site.

themselves (the newly reported star-shaped bacteria also show this characteristic). Possibly, this condition represents an artifact of fossilization, however, alternatively, the incised specimens may represent a strain of subspecies that are now extinct. Fossil populations may also be distinguished in having fewer individuals with a minimal number of rays (one to three) and more individuals retaining stipes.

C. Environmental Physiology

A few experimental studies have been undertaken to consolidate and clarify field-based data concerning microhabitats.
Evidence for temperature limits and tolerances has been based upon sampling in tropical alpine and subarctic thermal gradients together with a few laboratory variations in incubation temperature. Comparisons have involved widely separated geographical sources, hence they contain an additional variable.
The culture of different source materials using a standard protocol shows that the field-based conclusions are essentially accurate (Table 6) but that large quantitative differences are present among sample sites.
These observations support most strongly the concept that *Kakabekia* is a heat-sensitive organism with respectable cry-

TABLE 6
Low Temperature Tolerance in Kakabekia[a]

Incubation temperature (°C)	Alaska[b]	Number of Kakabekia/cm^2 Mauna Kea[b]	Iceland[c]
35	0	0	0
25	25	4	9
4	22	4	4
-7	9	>4	>4

[a]Based on 7-day cultures in routine ammonia water - glucose medium. Counts derived from sampling 10 - 15 fields on triplicate safranin-stained smears at 1250X (oil).

otolerance, rather than a true cryophile.

The question of alkaline requirements is perhaps more critical than thermal limits. Is ammonia required *per se* for culture, or can it be replaced by other alkaline media? Certainly there are many ammoniacal microhabitats; however, our routine culture medium offers no possible association with natural ammonia levels. A number of *Kakabekia* samples has now been grown in 0.01 *N* NaOH (ph 12) nearly as well as they were in 15*M* aqueous NH_3 (Table 7). It should be kept in mind, however, that the organic constituents of soil inocula could serve as a low level, but continuous, NH_3 generator under alkaline conditions.

The replacement of NaOH by sodium metasilicate provides nearly the same pH, but growth was quite diminished or completely suppressed. The exceptions were some of the samples that show poor growth under "normal" conditions. Although we have long associated silica with *Kakabekia* structure, this is the first indication of an adverse or toxic response to silicate. Presumably, the normal SiO_2 requirement of *Kakabekia* is met by the silica in the inorganic fractions of soil inocula themselves.

Since 1965, accumulated observations point to a correlation of some interest between the presence of diatoms or diatom wall fragments and the subsequent appearance of *Kakabekia* in culture.

Some 70 soil samples of geographically diverse origin were selected on the basis of the presence or absence of diatoms and the status of the culture with respect to *Kakabekia*. The results showed that though there was a number of samples with diatoms and the absence of *Kakabekia*, in no case did *Kakabekia* manifest itself without diatoms also being present in the inoculum. The two most obvious connections are that a common microenvironmental factor such as temperature or oxygen is required by both

TABLE 7

Comparative Performance of Kakabekia Soil Samples in Ammonia
Water and Other Alkaline Media at pH 12

Sample	Kakabekia sample density (N/cm^2)		
	NH_3 (15 M)	NaOH (0.01 M)	Na_2SiO_3 (0.01 M)
Alaskan Mendenhall Glac-ier, Juneau	21	17	9
Alaskan Point Barrow region	4	4	0
Reykjanes Peninsula, Ice-land	9	9	4
Mauna Kea (4000 m), Hawaii	4	4	4
Haleakala (2800 m), Maui	8	8	4
Himalayan foothills (3500 m).	present	absent	not tested
Near Mexico City (2000 m)	present	absent	not tested
Bogota, Columbia (2800 m)	present	absent	not tested
Quito, Equador (2500 m)	present	absent	not tested

organisms or that diatom silica is used by Kakabekia to meet
its own Sio_2 requirements.

The toxicity of complex organic substances was encountered
earlier during efforts (still unsucessful) to obtain Kakabekia
in pure culture. Metal ion toxicity was not considered until
geothermal mercury became an issue in Iceland.

Field collections in Iceland during June and July of 1970
and 1972 established a general distribution pattern for Kakabekia with 12 widely spaced positive sites out of 25. Upon
comparing the distribution patterns of Kakabekia and of soil
mercury, their inverse features were fully evident (Table 8).
Mercury toxicity, albeit at rather high levels, was demonstrated using two soil samples from South Iceland (Table 9). The
addition of fluorine, which is, like mercury, an emission product of Icelandic volcanic degassing process, failed both to
inhibit when added to the media and to alter the mercury response.

This chapter establishes that K. barghoorniana, a heat-
sensitive cryotolerant form, has a rather novel distribution
if one assumes an ancestral form living in silica-rich, possibly thermal, waters in a relatively cool environment.

Potentially, a great deal may be read into the morphological diversity of Kakabekia populations. Structural differentiation may appear more impressive and dynamic than is in fact
the case. Consider the fossil population, assuming the deeply

TABLE 8
A Comparison of Soil Mercury and Kakabekia Distribution Patterns
in Iceland

Location in Iceland	Mercury level (ug/kg dry soil)	Kakabekia
Flaajökull	3	present
Höfn	3	present
Vidarvatn	3	absent
Heimaey	5	absent
Surtsey	6	absent
Reykjavik	7	present
Hveragerdi	10	present
Myvatn	17	present
Skógar and Solheimajökull	31 - 42	present
Laugarvatn	35 - 80	present
Seliaheidi	68	present
Breidavik	70 - 200	absent
Hraunhafnartingi	100	absent
Hekla	400	absent

TABLE 9
Sensitivity of Kakabekia to Mercury and Floride

Inhibitor		Kakabekia sample density (N/cm^2)		
Hg^{+2} ppb	F^- ppm	Solheim glacier	Skogar Falls	Average
0	0	6	6	6.0
10	0	6	6	6.0
100	0	4	3	3.5
1000	0	0	0	0
0	10	5	6	5.5
0	100	5	5	5.0
0	1000	4	5	4.5
10	10	5	6	5.5
100	100	3	4	3.5

incised forms to reflect geologic modification of lobed or entire mantles, the actual extent of change in mantle structure
(even at its most extreme) has not so altered the population
or the individual that they are no longer recognizable. Hence,
the extent of change may in fact be remarkably modest for over
two aeons.

Acknowledgments

This work was carried out in collaboration with Dr. S. M. Siegel of the Department of Botany, University of Hawaii. Thanks are also due to the National Geographic Society, the Research Corporation (Cottrell Foundation), the National Aeronautics and Space Administration, and the University of Hawaii Research Foundation for their generous support.

References

Barghoorn, E. S., and Tyler, S. 1965. *Science, 147*:563.
Hirsch, P. 1974. *Ann. Rev. Microbiol., 28*:391
Nikitin, D. J. 1966. *New and Unusual Forms of Soil Microorganisma.* Isdat. Nauk, Moscow (in Russian).
Nikitin, D. J. 1973. *Bull. Ecol. Res. Comm.* (Stockholm), *17*:85.
Siegel, S. M., and Giumarro, C. 1966. *Proc. Natl. Acad. Sci. USA, 55*:349.
Siegel, S. M., Siegel, B. Z. Eshleman, A. M., and Bachman, K. 1973. *Environ. Biol. 2*:81.

16

A DISCUSSION OF BIOGENICITY CRITERIA IN A GEOLOGICAL CONTEXT WITH EXAMPLES FROM A VERY OLD GREENSTONE BELT, A LATE PRECAMBRIAN DEFORMED ZONE, AND TECTONIZED PHANEROZOIC ROCKS

M.D. MUIR, P.R. GRANT, G.M. BLISS, AND W.L. DIVER
Royal School of Mines, London

D.O. HALL
Kings College, London

In view of the continuing controversy about the biogenicity of structurally preserved carbonaceous remains in Archaean rocks, carefully selected samples from the Onverwacht Group, 3.355 X 10[9] years old, of South Africa, have been subjected to stringent morphological and statistical tests. The same tests have been applied to other samples from rocks of similar lithology that have undergone comparable geological histories from the Late Precambrian part of the Dalradian Supergroup and the Ordovician Ballantrae Group, both from Scotland. The morphological criteria applied are based on those outlined by Cloud and Licari (1968, Abstr. Geol. Soc. Amer. Ann. Meet. Mexico City), and the statistical tests are based on those formulated by Schopf (1975, Origins of Life, in press). 57 criteria used in this paper, assemblages from all three suites of rocks appear to be biogenic, and a consideration of the geological evidence supports this contention.

This chapter is intended to illustrate a rigorous geological, or, more specifically, micropaleontological, approach to the interpretation of chemically organic structures present in some ancient and metamorphosed sedimentary rocks. A stringent assessment of existing criteria for the determination of biogenicity of such microstructures is fundamental to such work. In this context it is perhaps useful to remember that the term fossil at one time carried no connotation of biogenicity, but was widely applied to any object "dug up" (from the Latin *fossilis*). Although the obvious similarity between some fossil objects and living organisms was early rec-

ognized, the concepts which rationalized such interpretations
did not develop until the 17th and 18th centuries (Rudwick,
1972). Well-preserved material was essential to the development of these concepts, and Precambrian micropaleontology
has had a comparable development. The mainstream of paleontological and biological science was unwilling to accept
microfossil evidence as a record of Precambrian life until
Tyler and Barghoorn (1954; Barghoorn and Tyler, 1965) presented
microfossils which were unequivocally both biogenic and Precambrian.

Many authors (see Schopf, 1975a, for a review) have
described morphologically distinctive microfossils from various horizons in the Proterozoic. The application of paleo-
botanical (paleobiological) methods to the study of these
microbiotas has greatly extended our knowledge. Notwithstand-
ing the small sample of Precambrian life that the described
microbiotas encompass, it is clear that morphologically distinctive microfossils become more elusive in progressively
older sedimentary rocks. This vexing problem is the result
of many factors, particularly the vagaries of preservation and
the apparently simple morphologies of the oldest Precambrian
microfossils.

The biogenicity of organic walled remains in the 3.355
x 10^9-year-old Onverwacht Group (Hurley *et al.*, 1972) of South
Africa has been regarded as dubious by a variety of workers
since they were first reported by Engel *et al.* (1968). There
were numerous reasons for this caution:

(1) The structures occur in rocks of great antiquity, and
at the time that Engel *et al.* wrote their initial report, very
few assemblages of Precambrian microfossils were known at all,
and only the well-preserved Gunflint chert microbiota (Barghoorn and Tyler, 1965) were at all well documented.

(2) The microstructures were not abundant and appeared to
be simple spheroids of a wide size range, with no clear modal
or mean size. It is implicitly assumed in nearly all discussions of this assemblage that because of the great age of
the host rock, the microfossils must belong to a single, more
or less homogenous population without any range of morphological
variability.

(3) In nearly all discussions of the assemblages, the "low"
metamorphic grade of the host rocks is stressed (see Engel
et al., 1968; Nagy and Nagy, 1969; Brooks and Muir, 1971;
Brooks *et al.*, 1973). The Onverwacht Group has been metamorphosed to greenschist grade, which involves thermal alterations
and development of new minerals, and this grade of metamorphism causes considerable alteration to the chemistry and morphology of organic matter.

(4) The stable isotope ratios of the Lower Onverwacht
Theespruit Formation are rather heavy ($\partial^{13}C \sim -16$ to -17%)

while the Upper Onverwacht values of $\partial \overset{13}{.} C \sim -24.9$ to -33.0% are more comparable with other Precambrian microfossil assemblages (Oehler et al., 1972). These heavy ratios have been interpreted as possibly indicating a not completely biogenic origin for the reduced carbon in the Theespruit Formation or possibly a derivation from nonphotosynthetic organisms. The values for the Upper Onverwacht Group are consistent with a biological origin for the reduced carbon.

(5)   It has been suggested by Folsome et al. (1975) that some of the spheroids from the Onverwacht Group samples may actually represent the remains of abiotically produced microspheres.

With these reservations in mind, we have been examining, over the last few years, cherts from the Onverwacht Group, both in thin sections and as disaggregated, macerated preparations (Brooks and Muir, 1971; Brooks et al., 1973; Muir and Hall, 1971; Muir and Grant, 1976), and in the present paper, we submit our data to a variety of tests to establish their biogenicity clearly and to disentangle the confusion that is threatening to clog the literature.

Assemblages from three suites of samples were selected for this comparative study. They were collected from the Onverwacht Group of South Africa, from the latest Precambrian of the Dalradian Supergroup of Argyll, Scotland, and from the Ordovician Ballantrae Group of Bennane Head, Ayrshire, Scotland. All of these deposits have undergone some degree of metamorphic and structural alteration. There are striking lithological similarities between the Onverwacht and Ballantrae sequences, and the Dalradian samples were selected as having undergone complex tectonic modifications.

The Onverwacht Group Material was collected by D. O. Hall from the Kromberg Formation, along the banks of the Komati River, at Skaapbrug, near the old JCI camp. Due to the large number of samples available to us, we have been able consistently to use the best preserved materials (Figs. 1 - 3) to provide the data used in statistical and morphological tests.

Most of the microfossils occur in bedded, fine-grained, black cherts, which are interbedded with volcanic and pyroclastic rocks. These cherts are microfractured in many places, and care was exercised in choosing samples for thin sections that were as little fractured as possible. In almost every case the microfractures were filled with coarser-grained, water-clear quartz. However, it is important to find areas in thin sections where there is little evidence of recrystallization. Microstructures in areas where recrystallization is suspected and where examination under crossed polars appears to indicate past grain boundaries which do not coincide with the present ones have not been considered in our statistical work. Furthermore, structures preserved in oxides or hydroxides of iron (Engel et al.,1968)are excluded from the present work.

Recent work on the black pyritic Easdale Slates of the
Middle Argyll Group at Easdale and on the more highly deformed
Tayvallich Slates from the top of the Argyll Group at Loch Nant
(Dalradian Supergroup, Harris and Pitcher, 1975, samples col-
lected by G. M. Bliss) have produced assemblages of microfossils
(Figs. 4 and 5) from rocks metamorphosed to the greenschist
grade that have undergone at least two stages of deformation,
reorienting the phyllosilicate minerals and mobilizing the quartz
(Bliss, in preparation).

The Ordovician sample was collected by W. L. Diver from
the Bennane Burn, south of Bennane Head (see Fig. 4 and Bailey
and McCallien, 1957). The lavas and cherts of this sequence are
less metamorphosed than either the Onverwacht or the Dalradian
rocks, but the cherts contain rather similar spheroidal microfossils (Figs. 6 - 8). The bedding structures and distribution

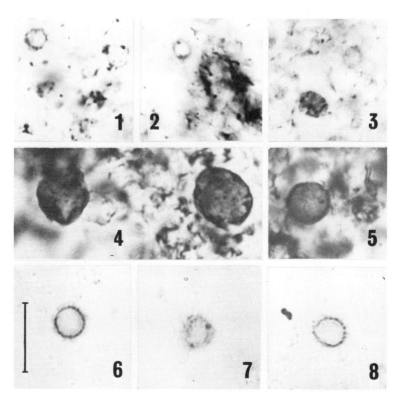

*Figs. 1 - 3. Small spheroids from the Kromberg Chert, Ko-
mati River, South Africa. Bar marker = 10 μm. Petrographic thin
section.*

*Figs. 4 and 5. Small spheroids from the Easdale Slates,
Argyll, Scotland. Petrographic thin section.*

*Figs. 6 - 8. Small spheroids from the Ballantrae Formation,
Bennane Head, Scotland.*

of the organic matter in these Ordovician cherts are strikingly
similar to those in the Kromberg Formation samples.

Throughout this study, normal 30-μm-thick, petrographic
thin sections were used. Measurements for the statistical work
were made using an eyepiece micrometer, directly, in the optical
microscope, in the case of the Ordovician samples. For the Dalradian and Onverwacht data, photomicrographs of thin sections of
fossiliferous rock were taken at a magnification of x400 on the
negative, using a x100 oil immersion lens. The negatives were
enlarged to a consistent size (x4000) in a photographic enlarger,
and measurements were made using a transparent millimeter grid,
accurate to 1 mm.

The spheroids in all three sections are brown or black.
A variety of morphological types can be distinguished in each
of the samples. Muir and Hall (1974) and Muir and Grant (1976)
have described several types of spheroids, filaments, and even
colonial structures in the Onverwacht Group, and these have also
been observed in the present study. The state of preservation
of the microfossils is rather poor, and the organic matter has
frequently been rearranged by recrystallization of mineral matter. Any specimens that had been affected by such recrystallization or that were coincident with mineral grain boundaries were
scrupulously excluded from the measurements.

At least six morphologically recognizable microfossil
types, generally well preserved, have been found in the Dalradian
Easdale Slates (Bliss, in preparation). In the more highly deformed samples (i.e., where there is development of an intense
secondary crenulation cleavage), the fossils are very dark,
the wall may have become punctured and degraded. Walls may often remain intact, however, and morphological features, such as
spines, may also be preserved. These features are also found in
assemblages from the more highly deformed Dalradian Tayvallich
Slates. Preservation of microfossils in these rocks has been
achieved by encapsulation of these fossils within mobilized
quartz, which has acted subsequently as a relatively protective
microenvironment. Downie et al. (1971) have been able to provide invaluable biostratigraphic data from their palynological
examination of the Dalradian Supergroup.

The Ordovician rocks contain a number of types of sphero-
morphid Acritarchs (Downie, 1973), some of which are spiny
(Acanthomorphitae) and some of which are smooth (Sphaeromorphitae)
For the purposes of this paper, one morphological type of spher-
omorph only was measured.

In all samples used for this study, only spheroids with a
repeated and consistent morphology were measured. A number of
statistical tests have been applied to the simple spheres of the
assemblages in the rocks, including tests for normality of distribution (see Muir and Grant, 1976) and three tests suggested
by Schopf (1975b): a size/probability plot, a divisional dis-

persion index (DDI), and a standard deviation. These are outlined below:

(1) The fit of the assemblages to a normal distribution,
or combination of such distributions, by means of a probability
plot.

(2) The divisional dispersion index (DDI), which is the
least number of steps necessary to reduce the size of the largest
individual to that of the smallest in a monotypic assemblage, by
a series of "mathematical vegetative (i.e., asexual) divisions."
Schopf is of the opinion that for an assemblage to be biogenic,
the DDI should fall within the range of about 3 to 8.

(3) The standard deviation of the size distribution expressed as a percentage. Schopf suggests that for a biogenic
assemblage, this should be low to moderate (i.e., 12 - 24%).

According to Schopf (1975b), a monospecific biogenic assemblage will approach a straight line on the probability plot
(see Fig. 9), similar to those of species of the modern algal

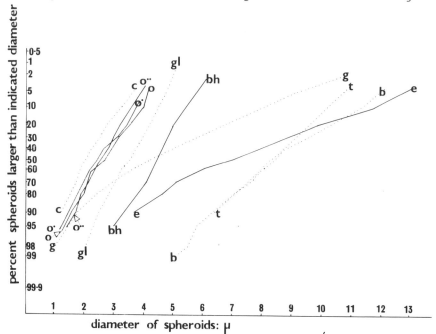

*Fig. 9. Hyperbolic probability/size plot [after the same
style as those plotted in Schopf (1975b)]. c = Chlorella, gl =
Gloeocapsa, g = Gunflint, t = Tetraspora, b = Bitter Springs
"spot cells." These plots (dotted lines) are redrawn from Schopf
(1975b). o = Bedding plane 1, o· = Bedding plane 2, o·· = Bedding plane 3. These plots are from the Kromberg chert (raw data
are given in Muir and Grant, 1976). bh = data for the Ordovician
Bennane Head Ballantrae Group sample. e = data for small spheroids
from the Easdale Slates of the Dalradian.*

genera *Chlorella, Gloeocapsa,* and *Tetraspora,* while more complex
heterogeneous assemblages should be similar to the plots of the
undoubtedly biogenic Gunflint unicells, and the Bitter Springs
"spot cells." When the data for the assemblages in three bedding planes from the Kromberg Formation chert (see Muir and Grant,
1976 , for further details) and the data for the Ordovician
spheroids are plotted in the same manner, they form a tight group
of very nearly straight subparallel lines close to the plots produced from the modern algae. The plot of the Easdale Slates,
however, lies parallel and just "below" that of the Gunflint unicells. This suggests that the style of the Bennane Head assemblage is similar to that of the three bedding plane assemblages
of the Kromberg Formation, and that all the assemblages are simple and probably monospecific. The complex Easdale Slates assemblage is similar in style to the complex assemblage of Gunflint
unicells. The DDIs from all of these assemblages fall within
Schopf's limits for biogenicity--Bennane Head DDI = 4; Onverwacht
DDI = 7; Easdale Slates DDI = 7--and the standard deviations also
fit Schopf's limits because the parameters are mathematically
related. The results of these statistical tests demonstrate
that within the limits imposed by Schopf (1975b), these assemblages are all biogenic. It should be remembered, however, that
this in no way implies (a) any suggestion of biological affinities
nor (b) that the assemblages bear any biological relationship to
one another. The results simply indicate that the statistical
compositions of the simple and complex groups of assemblages are
similar.

Apart from the statistical tests outlined and discussed
above, the biogenicity of the structures in all three assemblages
is based on a number of widely applicable morphological criteria
which are largely an extension and elaboration of those first
advocated by Cloud and Licari (1968). One of the major advantages of using well-founded morphological criteria is that each
can be tested under a good polarizing microscope without recourse
to sophisticated and sometimes misleading chemical or statistical
techniques.

The morphological criteria used are as follows:

(1) The microstructure should be of similar color to the
surrounding organic debris. If this is so, then it is likely
that the microstructures and the organic debris have undergone
a similar geological history. It must be remembered that dense
clumps of organic debris may appear unusually dark, as may especially thick-walled microstructures.

(2) The microstructures should be of distinctive and consistent morphology and they should occur abundantly throughout
a sequence of rocks. Simple spheroids present special difficulties, but when dealing with Precambrian rocks, they are common. Distinctive morphologies that appear only very rarely
should be treated with circumspection.

(3)   The microstructure should be demonstrably of organic
matter. This can be satisfactorily determined by means of the
optical properties, using a polarizing microscope, but should
be confirmed chemically (e.g., by using X-ray diffraction.
(4) There should be no mineralogical distortion of the microstructure. The growth of disruptive minerals can begin during
or after deposition,and can occur as a response to diagenesis
and metamorphism any number of times during the rock's history.
New crystal growth can distort the microorganism. However,
crystal growth need not necessarily distort the microstructure,
as for example, in the case of permineralization and/or replacement.
(5)   The microstructure should occur in what is demonstrably a rock of sedimentary origin and be syndepositional
with the sediment. These found in postdepositional cracks and
joints should be treated with reserve. The rock should be
fresh when collected in the field and away from both present and
paleoweathering surfaces.

After careful examination of the organic matter in all
the samples, all the above criteria appear to be fulfilled by
the small spheroids from all the rocks used in this study.
Statistically and morphologically, therefore, the spheroids in
all three samples appear to be biogenic.
We must now consider the other arguments which, in particular, concern the organic walled microstructures in the Onverwacht Group.

(1)   The extreme age of the rocks in no way precludes
the possibility of microfossils occurring within them. The
known occurrences of stromatolites are a part of the gradually
accruing body of acceptable evidence for Archaean (>2.5 x 10^9
years old) biogenic activity. Such stromatolites have been
found in suitable depositional environments that have been
fortuitously preserved from excessive tectonism, erosion, or
burial during subsequent geological history. Figure 10 shows
the world-wide distribution of these occurrences, only two of
which were known (MacGregor, 1941; Walcott, 1912) at the time
of the first report of organic microstructures in the Onverwacht.

Although there have been no reports so far of microfossils
in the banded iron formations (BIF), which commonly occur in
the Archaean, examples abound in Proterozoic BIFs (e.g., Gunflint chert assemblages in Gunflint BIF, Barghoorn and Tyler,
1965), and BIFs are regarded by many authors as containing
evidence for biochemical activity (Tendall and Blockley, 1970).
The graphite described from very ancient (3.76 X 10^9 years
old, from Isua, Greenland) Archaean rocks by Nagy *et al.* (1975)
occurs in a vein in a micaceous metaquartzite and is not, therefore, syndepositional with the metaquartzite. By our above criteria, this occurrence does not warrant consideration as evidence

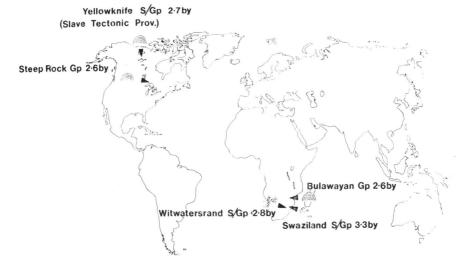

Fig. 10. Distribution of reported fossils from rocks >2.5
x 10⁹ years old. Yellowknife Supergroup (Henderson, 1974),
Steep Rock Group (Walcott, 1912), Bulawayan Group (MacGregor,
1941), and Witwatersrand Supergroup (Hallbauer, 1975). For
Swaziland Supergroup, see reference list, especially Schopf
(1975b). In this diagram, the small, stacked hemispheres rep-
resent stromatolite occurences, the tubes represent filamentous
forms, and the small spheroids represent other microfossils.

of biogenic activity in the Isua specimen. Consequently, the contention of Nagy et al. (1975) that a nonbiogenic origin for these
graphitic microstructures can be adduced to question the biogenicity of the demonstrably indigenous microstructures of other Archaean rocks (i.e., Onverwacht and Fig Tree Groups) is, in our
view, indefensible.

    (2)   A number of workers has reported wide size ranges for
the Onverwacht Group microfossils (Engel et al., 1968; Nagy
and Nagy, 1969) and this has been dealt with in detail by
Schopf (1975b). A simple explanation of wide size ranges is
probably that the assemblage is not, as is implicitly assumed
by many workers, monotypic, but represents a variety of morphological entities. It is possible that some of the very large
spheres described by Engel et al. (1968) may actually be associated with colonial structures. A certain amount of original
morphological variation has been demonstrated by Muir and Hall
(1974) and Muir and Grant (1976), but the microfossils have
undergone numerous physical and chemical changes resulting from
processes of preservation and metamorphism.

    (3)   Although depositional and early diagenetic effects
(e.g., permineralization) may protect a potential fossil in a
particular matrix, the geological factors of burial, tectonism,

and chemically active pore fluids in the phreatic and vadose
zones compound with time to minimize the chances of fossils
remaining in an unaltered matrix. In addition, replacement
and recrystallization of the matrix almost invariably further
physically modify or damage organic walled fossils. Considering these geological factors, it is highly probable that
both physical and chemical alteration and degradation will
occur and that this modification will drastically alter the
morphological details preserved. The factors combine to reduce
the chances of finding distinctive fossils in very old and/or
metamorphosed rocks.

Organic matter alters its color and chemical composition
during thermal metamorphism (coalification). Gutjahr (1966)
showed experimentally that pollen of *Quercus robur* changes its
color from greenish yellow (when fresh) to dark brown at about
250°C. Sengupta (1974, 1975) showed that *Lycopodium clavatum*
spores respond in a similar way, becoming dark brown at 250°C
and black at 350°C (Figs. 11 and 12).

She was also able to demonstrate that other spores and
pollen grains darken with increasing temperature (Sengupta,
1976) and that the wall of the palynomorphs is physically and
structurally modified as a result of heating and pressure
(Sengupta and Rowley, 1975). The color changes of spores and
pollen grains and other palynomorphs are used in exploration
for oil and gas to determine paleotemperatures of sediments
(Staplin, 1969), and Correia (1971) was able to show that
different groups of organic walled microfossils respond more or
less gradually to temperature increases--e.g., spores and
pollen grains darken more rapidly than dinoflagellates. This
variability of response presumably reflects initial chemical
differences. Gray and Boucot (1975) have demonstrated that
shear stress can also cause coalification of microfossils.

Precambrian organic walled microfossils are susceptible to
similar metamorphic changes. Cells from the Bitter Springs
Formation of Australia (see Schopf, 1968) are light to midbrown
in color and occur in an unmetamorphosed sequence. However,
cells from the probably correlative "Skillogalee Dolomite" of
Boorthanna, South Australia (Schopf and Fairchild, 1973), which
has undergone greater metamorphism, are black or very dark
brown, and while the overall structure of the cells remains
recognizable, the fine structure of the walls has been consid-
erably rearranged (Figs. 13 and 14). The Skillogalee Dolomite
is not highly metamorphosed in classic Barrovian terms. It
does not reach greenschist facies, and the Dalradian and
Onverwacht rocks with which we are concerned are more metamorphosed even than these. Thus, it is reasonable to expect that
the organic matter of the microfossil walls will be black and
will have suffered some structural modification, both as a
direct result of the metamorphic temperature increase and as
a result of recrystallization of the minerals of the rocks in

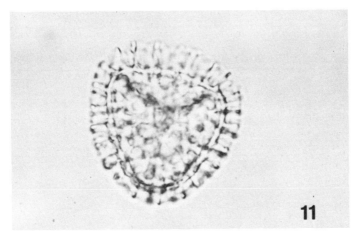

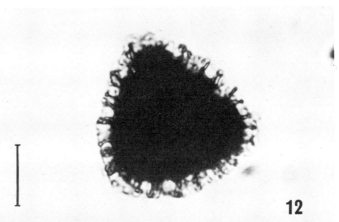

Fig. 11 and 12. Lycopodium clavatum *L. spores. Figure 11 is fresh, untreated material, Figure 12 has been experimentally heated to 250°C for 100 hr. Bar marker = 40 µm.*

which the microfossils occur. Therefore, preservation of fine details [such as can be found in the Bitter Springs Formation (Schopf, 1968; Oehler, 1976), the McArthur Group (Croxford *et al.*, 1973; Muir, 1974, 1976), and the Bungle Bungle Dolomite (Diver, 1974)] cannot be expected. Within the metamorphic limitations, however, microfossils may be preserved reasonably well even though they have undergone chemical and structural alterations, and these can be identified by experienced workers.

(4) Thus, morphologically, we can identify metamorphic changes in organic matter in the cherts of the Onverwacht Group, and this has been confirmed by Karkhanis (1975), who

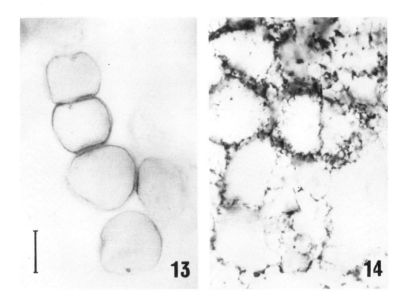

Figs. 13 and 14. Figure 13 shows a group of cells from the more or less unmetamorphosed Bitter Springs Formation. Figure 14 shows a group of cells from the more highly metamorphosed "Skillogalee Dolomite" of Roorthanna. Bar marker = 10 μm. Petrographic thin sections.

was able to demonstrate the presence of graphite in these samples. Futhermore, McKirdy and Powell (1974) showed that thermal metamorphism affects $\delta^{13}C$ values by driving off ^{12}C, leaving the residue enriched with ^{13}C. Thus the "anomalously" heavy $\delta^{13}C$ values obtained from Lower Onverwacht samples can be explained as a result of metamorphism, either regional or local thermal effects from extrusive lavas (Muir and Grant, 1976).

(5) While Folsome *et al.* (1975) have been able to demonstrate that organic microstructures--some of which are similar to life - like forms--may be produced by abiotic syntheses, the probability of such processes being responsible for the widespread occurrence of the Onverwacht and Fig Tree spheroids (Schopf, 1975b) is not consistent with the paleoenvironments at these times suggested by Anhauesser (1973). Furthermore, the credibility for such a mechanism is diminished by the fact that the unaltered synthetic products do not resemble living organisms but metamorphosed relics.

Evidence from organic microsturctures in samples from the 3.355×10^9-year-old Kromberg Formation of the Onverwacht Group has been subjected to morphological and statistical tests proposed by others as criteria for evaluating the biogenicity of

organic microstructures occurring in sedimentary rocks. In all
cases, our Kromberg data are consistent with the morphological
criteria and meet the statistical requirements. Hence,
on these bases, the microstructures may be considered to be
biogenic and thus may justifiably be called microfossils, since
they are judged to be a record of past life. Similar evidence
has been obtained from the younger Dalradian and Ordovician
microfossiliferous formations which (a) have been subjected
to a comparable grade of metamorphism, and (b) in the case of
the Ordovician samples, are from a similar sequence of pillow
lavas, pyroclastic sediments, and shales and bedded cherts.
The last of these lithologies has striking lithological and
petrographic similarities with the fossiliferous Kromberg
cherts, but is less metamorphosed. The data from these samples
have been subjected to similar scrutiny as for the Kromberg
data and, not surprisingly, have also yielded positive results.

The study of the origins and nature of early life forms
on earth has become an important unifying theme between various
branches of chemistry, biology, and geology. The interdisciplinary nature of much of the research on the Precambrian makes
it a most stimulating arena for scientific activity. However,
while the interchange of ideas and methodologies between disi-
plines is exciting and fruitful, it should be remembered that
each discipline has certain intrinsic subtleties. Such nuances
are not swiftly culled from the literature nor easily absorbed
from lectures and discussions at meetings, but are the products
of experience and individual judgement. The nature of geological evidence often places *a priori* constraints on its interpretation, and much of it is, at best, equivocal and often circumstantial. This is particularly true of paleontology, where
intuition and experience are inextricably combined with objective
rigor.

Thus, in addition to the rather simplistic approaches outlined above, we have adduced what we consider to be relevant
geological evidence, which may be briefly summarized. The
abundance of organic matter in our Kromberg chert samples, and
its textures and distribution within the least-altered parts
of the rock, are very similar to those from unequivically
fossiliferous (i.e., containing acritarchs and radiolaria)
Ordovician cherts from a very similar geological setting. Organic microstructures of a variety of morphologies similar to
those of known simple organisms and microfossils are common
in some of the Kromberg cherts. The microfossils from the Dalradian slates demonstrate clearly that microfossils are able to
survive the effects of metamorphism and structural deformation
(in this case, far in excess of that experienced by the Kromberg Formation).

On the basis of the type of reasoning outlined above, and
the evidence presented in this and two other papers (Muir and

Hall, 1974; Muir and Grant, 1976), the Kromberg Formation microstructures are considered to be the remains, albeit modified
by geological processes, of once living organisms. The nature
of the evidence in no way permits any suggestion of the kind
of biological affinity that should be assigned.

Thus, on the basis of two separate kinds of evidence, the
case for the biogenic origin of the Kromberg microstructures
in presented and, in our view, supported. Furthermore, it is
our contention that with careful fieldwork and preparation of
samples, more and better-preserved microfossils remain to be
discovered within the Onverwacht Group and other Archaean
sediments.

In conclusion:

(1) Precambrian microfossils ought to be obtained from
"sensible" (sedimentary) rock types.

(2) The likelihood of contamination or unecessary alteration should be minimized by the collection of fresh samples,
preferably obtained from well below present or paleoweathering
zones.

(3) Once thin sections and/or chemical residues are prepared, Cloud and Licari's (1968) criteria may be applied with
whatever level of extra sophistication is available: for example,
checking the color of the total organic content for contamination or reworking. It is particularly important when working
with chemical preparations to recognize both contaminants and
artifacts.

(4) It is imperative that microfossils should be demonstrably syndepositional with the rock. Clearly, they should be
rejected if there is any suspicion that they occur in veins,
cracks, or joints, or appear to be part of a replacement
cycle, or in the case of filamentous forms, if there is a possibility that they may be younger boring algae or fungi.

(5) When appropriate, statistical tests may be applied,
but the results of such tests must be considered as indicators
only, not as infallible proofs of biogenicity. The results are
not evidence in their own right, but require careful interpretation; For example, the hyperbolic probability/size plots
(Schopf, 1975b) may well be the result of the combination of
several statistical and/or biological populations. Deconvo-
lution of such curves may well be valid statistically, but has
no biological meaning. Thus, optical reexamination of the
assemblage should be made if multiple populations are indicated
by the statistical test. Forcing of the assemblages into
statistically indicated groups should not occur.

(6) Microfossil evidence should be accurately placed in
its geological context at the local and broader levels. Failure
to utilize classic geological methods is a waste of potential
information. The fragmentary nature of the Archaean record
does not permit the wastage of information.

Acknowledgments
We wish to thank the Natural Environment Research Council
for financial support for W. L. Diver and G. M. Bliss. P. R.
Grant thanks the Central Research Fund of London University
for travel funds. We are indebted to Dr. S. Sengupta for the
micrographs shown in Figs. 11 and 12 and to her and to Mr. S.
N. Karkhanis for helpful discussions.

REFERENCES

Annhaeusser, C. R. 1973. *Phil. Trans. Roy. Soc. Lond. Ser. A,*
273:359.
Bailey, E. B. and McCallien, W. J. 1957. *Trans. Edinburgh Geol.*
Soc., *17*:33.
Barghoorn, E. S., and Tyler, S. A. 1965. *Science, 147*:563.
Bliss, G. M., in preparation.
Brooks, J., and Muir, M. D. 1971. *Grana, 11*:9.
Brooks, J., Muir, M. D., and Shaw, G. 1973. *Nature (London)*
244:215
Cloud, P. E. Jr., and Licari, G. R. 1968. *Abst. Geol. Soc. Amer.*
Ann. Meet. Mexico City, 57.
Correia, M. 1971. In J. Brooks, R. P. Grant, M. D. Muir, P. van Gijzel
and G. Shaw. (eds.), *Sporopollenin.* Academic Press, London.
Croxford, N. J. W., Janecek, J., Muir, M. D., and Plumb, K. A.
1973. *Nature (London),* 245 (5419):28.
Diver, W. L. 1974. *Nature (London),* 247 *(5440)*:361.
Downie, C. 1973. *Palaeontology,* 16:239.
Downie, C., Lister, T. R., Harris, A. L., and Fettes, D. J. 1971.
In *A Palynological Investigation of the Dalradian Rocks of*
Scotland. Rep no. 71/9 Inst. Geol. Sci.
Engel, A. E. J., Nagy, B., Nagy, L. A., Engel, C. G., Kremp,
G. O. W., and Drew, C. M. 1968. *Science, 161*:1005.
Folsome, C. E., Allen, R. D., and Ichinose, N. K. 1975 *Precam-*
brian Res., 2:263.
Gray, J., and Boucot, A. J. 1975 *Bull. Geol. Soc, Amer., 86*:1019.
Gutjahr, C. C. M. 1966. *Leidse Geol. Meded., 38*:1.
Hallbauer, D. K. 1975. *Miner. Sci. En.* 7:111.
Harris, A. L., and Pitcher, W. S. 1975. In A. L. Harris, R. M.
Shackleton, J. Watson, C. Downie, W. B. Harland, and S.
Moorbath (eds.), *Precambrian: A Correlation of Precambrian*
Rocks in the British Isles.
Henderson, J. B. 1974. *Geol. Surv. Canad,* Paper 75-1, Part A:325.
Hurley, P. M., Pinson, W. H., Nagy, B., and Teska, T. M. 1972.
Earth Planet. Sci. Lett., 14:360.
Karkhanis, S. N. 1975. *Chem. Geol., 16*:233.
Macgregor, A. M. 1941. *Trans. Geol. Soc. S. Afr. 43*:9.
McKirdy, D. M., and Powell, T. C. 1974. *Geology,* p. 591
Muir, M. D. 1974. *Origins of Life, 5:105.*

Muir, M. D. 1976. *Alcheringa,* in press.
Muir, M. D., and Grant, P. R. In B. F. Windley (ed.), *Early History of the Earth.* Wiley, London.
Muir, M. D., and Hall, D. O. 1974. *Nature (London), 252:*376.
Nagy, B., and Nagy, L. A. 1969. *Nature (London) 223:*1226.
Nagy, B., Zumberge, J. E., and Nagy, L. A. 1975. *Proc. Natl. Acad. Sci. USA, 72:*1206.
Oehler, D. Z., Schopf, J. W., and Kvenvolden, K. A. 1972. *Science, 175:*1246.
Oehler, D. Z. 1976. *J. Paleontol. 50:*P. 90.
Rudwick, M. J. S. 1972. In *The Meaning of Fossils. Episodes in the History of Palaeontology,* Pp. 1 - 287. MacDonald and Co., Ltd., London.
Schopf, J. W. 1968. *J. Paleontol. 42:*651.
Schopf, J. W. 1975a. *Amer. Rev. Earth. Planet. Sci., 3:*213.
Schopf, J. W. 1975b. *Origins Life,* in press.
Schopf, J. W., and Fairchild, T. 1973. *Nature (London), 242:*537.
Sengupta, S. 1974. O. Johari and I. Corvin (eds.), *Procedures of Scanning Electron Microscopy,* 381 - 388. IITRI.
Sengupta, S. 1975. In *Actes 6eme Congr. Int. Geochimie Organique Technip, pp. 493 - 494.* Paris.
Sengupta, S. 1976. *Proc. Roy. Microscop. Soc.,* in press.
Sengupta, S. and Rowley, J. R. 1975. *Grana, 14:* 143.
Staplin, F. L. 1969. *Bull. Canad. Pet. Geol., 17*(1):47.
Trendall, A. F. and Blockley, J. 1970. *The Iron Formations of the Precambrian Hammersley Group, Western Australia, Bull. 119. Geological Survey of Western Australia.*
Tyler, S. A., and Barghoorn, E. S. 1954. *Science, 119:*606.
Walcott, C. D. 1912. *Geol. Surv. Canad. Mem., 28:*16.

17

SYNTHETIC ORGANIC MICROSTRUCTURES AS MODEL SYSTEMS FOR EARLY PROTOBIONTS

CLAIRE FOLSOME

University of Hawaii, Honolulu

The establishment of the earliest date in the Precambrian at which unambiguous living forms existed might prove to be a search guided more by our own conceptions of life than we suspect. Perhaps the Bulawayan stromatolites, 2800 - 2500 million years old (Schopf *et al.*, 1971; MacGregor, 1941) are indicative of the earliest well-evolved microbial forms. Other suggestive evidences of Archean life are rod-shaped organic bacterial-like bodies (Barghoorn and Schopf, 1966), filamentous microfossils (Brooks *et al.*, 1973), and spheroidal microfossils from the Swaziland sequence (see Schopf, 1975). However, these structures are difficult to interpret unambiguously as organic remains of living microorganisms.

One might expect that the earliest forms of microbial life from ~3800 million years (the age of the oldest known rocks of sedimentary origin) to ~2800 million years old (the Bulawayan stromatolites) would demonstrate less morphological complexity and more heterogeneity as expansion into many open ecological niches occurred and as early evolution and genetic evolution of metabolic pathways developed. It is conceivable that early protobionts lacked a genetic code (Fox, 1973). Early Archean microfossils are simpler in structure, more varied in size and morphology, and less convincing as to biogenicity than are more recent microfossils. Indeed, some microstructures of the South African Onverwach (ca. 3400 million years old) and Fig Tree Series (ca. 3000 million years old) closely resemble graphitic microstructures from the Isua micaceous metaquartzites ($376° ± 70$ million years old), which seem to be of abiotic origin (Nagy *et al.*, 1975).

Kerogen from unmetamorphosed or slightly metamorphosed
Archean sediments over 3000 million years old has a carbon isotope composition similar to that of geologically younger biologic matter (Schopf *et al.*, 1971). These data usually are taken as an indication of biogenicity. But, carbon isotope ratio
disproportionation is, in fact, indicative of nonequilibrium
or metastable equilibrium processes. Any nonequilibrium synthesis of organic matter will also show similar disproportionation as has been shown by Studier *et al.*, (1968).

As direct evidence of Archean life one is left with the
occurrence of Archean stromatolites and with the inference that
earlier Archean evolution (proceeding at an unknown rate) must
have occurred to have lead to forms capable of leaving stromatolites. Questions remain. How "biological" were these early
Archean predecessors? By what criteria, if any, does one distinguish a pseudofossil from an early biont? Can one construct
synthetic structures which model protobionts from the far side
of the Archean?

By chance, it recently was noticed that when an electric
discharge was passed through CO (or CH_4) and N_2 over a water
surface in a modified Urey - Miller experiment, the water soon
became turbid. Upon microscopic examination, numerous organic
microspheres were seen (Folsome *et al.*, 1975; Fraser and Folsome,
1975). Many of the properties of these structures seemed closely to resemble Archean microfossils. Studies were begun with
the object of gaining a more thorough comprehension of that now
vague transitional stage from chemical to biological evolution
during the origin of protobionts.

I. SYNTHESIS OF ORGANIC MICROSTRUCTURES

Microstructures were synthesized in a 2-liter flask containing 20 ml of double-distilled water degassed and filled with
200 mm of CH_4, 200 mm of CO (or CO_2), and 50 mm of N_2. A Tesla
coil was discharged continuously for periods of up to 3 days
through a gold or nichrome central electrode held 5 - 7 mm over
the water surface.

Several hours after sparking, particulate matter and turbidity were noted. After 1 - 3 days, the turbid yellow-brown
aqueous solution was removed and examined directly as water
mounts (for light microscopy), or sedimentable material was
collected and washed in double-distilled water for chemical or
other studies. About 60% of the input carbon is recovered in
these structures, measured by use of ^{14}C-labeled methane, carbon dioxide, or carbon monoxide.

II. MORPHOLOGY BY LIGHT MICROSCOPY

A survey of wet-mounted samples from several spark-discharge experiments was conducted using phase-contrast microscopy and Zeiss Nomarski differential interference phase-contrast microscopy. Particle counts were obtained using a Petroff - Hausser counting chamber. These data are summarized in Table 1. Figs. 1 and 2 are optical photomicrographs of representative structures.

TABLE 1
Classification or Organic Microspheres[a]

Type	Outline	Surface	Size (μm)	Abundance (per milliter)
I	Spheroid	Rugose and textured	20 - 40	5.5×10^5
II	Spheroid	few marked folds	10 - 15	3×10^6
III	Bacterial-like spheres and rods	Smooth	1 - 2 in diameter	8×10^7
IV	Tubular	Smooth-fibrous hollow centers	4 - 8 x 70	5×10^5
V	Massive complexes	--	150 x 60, or greater	2×10^5

[a]Data for this survey were taken using Zeiss phase-contrast optics at 450x (Folsome *et al.*, 1975).

A survey of wet-mounted samples from several spark-discharge experiments was conducted using phase-contrast microscopy and Zeiss Nomarski differential interference phase-contrast microscopy. Particle counts were obtained using a Petroff - Hausser counting chamber. These data are summarized in Table 1. Figs. 1 and 2 are optical photomicrographs of representative structures.

It appeared possible to "classify" the assembly into types based upon size, outline, and surface morphology (see Table 1). Large (20 - 40 μm) and intermediate (10 - 15 μm) spheroids are most evident, but most common were bacterial-like rods (1 x 2 - 3 μm) and spheres (1 - 2 μm). Tubular structures of various diameters and lengths (4 - 15 x 20 - 80 μm) as well as massive complexes of tubes and sheets (150 x 60 μm) were also evident.

Large and intermidiate spherical elements frequently appear as clusters, in some cases associated with enveloping membranous material. Nomarski optics show more clearly that the

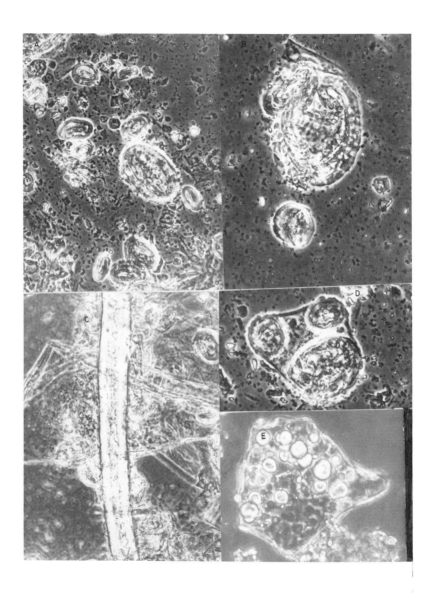

Fig. 1 *Phase-contrast photomicrographs of organic micro-structures as water mounts, undiluted. A and B: types I and II spheroids. The larger type I's are about 20 - 25 μm in diameter. D and E: type II spheroids 10 - 20 μm in diameter, clustered within a surrounding membrane-like material. C: tubular material within a clump of type II spheroids. Note the great numerical majority of 1 - 2-μm bacterial-like structures throughout the background of all pictures.*

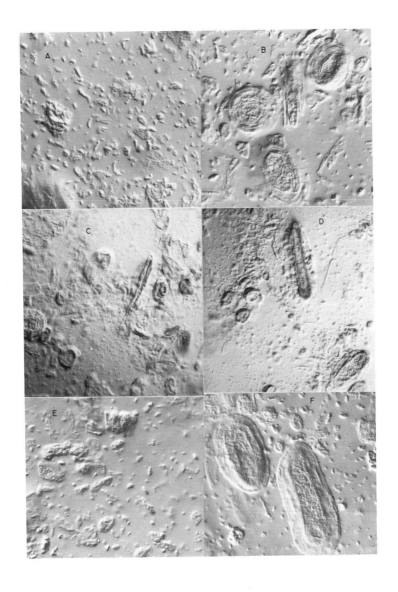

Fig. 2 Zeiss - Nomarski differential interference contrast photomicrographs of water mounts of organic microstructures. Note the smooth outer (?membranous) surfaces, as in F. B,C,and D, taken at a lower focal plane, show a quasi-crystallinic regularity of inner particulate units. Fragments of broken tubular structures are present in B,C,and D while clumps of type III small bacterial-like structures are present throughout, but are most evident in A.

outer surface of these structures is continuous and membrane-
like.

III. MORPHOLOGY BY TRANSMISSION ELECTRON MICROSCOPY

Aliquots of a diluted sample of washed organic microstructures were dropped onto Formvar grids and, after drying, were
examined unstained and unfixed by transmission electron microscopy (see Folsome et al., 1975). Specimens were of low electron density and appeared as a gossamer network of filamentous
and membranous matter. The bacterial-like smaller structures
were more obviously composed of aggregates of particles 0.028
μm in size. Similar particles are seen enmeshed in a sheet-
like membranous matrix for the larger spheroids.

IV. CHEMICAL STUDIES

The structures can be centrifuged from aqueous solutions,
and all but the larger tubular elements retain their form. This
implies that their shapes are stable ones and that densities
exceed 1 g cm^{-3}. Upon dehydration, the structures collapse,
but they retain their morphology when rehydrated.
Water-washed dried structures were ground with KBr and
examined for infrared absorption from 2 to 16 μm: maxima at 2.92
μm, 5.8 - 6.3 μm, and a broad band from 8.5 - 10.1 μm were recorded.
After prolonged 6 N HCL (120°C) hydrolysis, much of the material remained sedimentable. The soluble organic fraction,
when separated by thin-layer chromatography, showed a variety
of ninhydrin positive material (amino acids) and a hint of chro-
maphores.
The material behaves as a simple form of "kerogen" which
appears to contain extensively cross-linked polypeptides. Work
is still in progress to elucidate the chemical composition of
these structures.

V. CATALYTIC PROPERTIES

Organic microspheres were prepared essentially as described
above, but strict aseptic techniques were employed throughout
(e.g. autoclaving and dry air sterilizing all glassware--using
autoclaved and filtered double-distilled water). Microspheres
were harvested by centrifugation, concentrated 10-fold, and
tested for ATPase and H_2O_2 decomposing activities. ATPase
activity was measured by release of orthophosphate and H_2O_2 decomposition was determined by use of an Orion oxygen electrode.
Sterility was checked by incubating an aliquot of microspheres
in sterile Brain Heart Infusion broth at 23°C for 1 week.

Faint activities were noticed for both assays in early experiments. However, long-term incubation of these microsphere preparations revealed a slow-growing fungal contaminant. No activities were detectable above background when uncontaminated samples were used.

VI. KINETICS OF SYNTHESIS

Periodically, aqueous samples were withdrawn from a continuously sparked flask containing 300 mm of CH_4, 100 mm of CO, 50 mm of N_2, and 50 ml of water. These aliquots were examined and counted using a phase-contrast microscope at 900x and a Petroff-Hausser counting chamber. These experiments are reported in Fraser and Folsome (1975).

Exponential kinetics of synthesis were observed for all experiments and for all major spheroidal morphological types after a lag of about 10 hr. At about 30 - 40 hr the rate of synthesis decreases, at which time the particle count reached a maximum. The slope of the exponential portion of the curves for these three experiments gave particle doubling times of 9.5, 6.5, and 4.0 hr. When the telsa coil was turned off during the exponential period, particle production ceased, but resumed upon continued sparking.

VII. DISCUSSION

High yields of discrete groups of structured entities are formed by the action of a spark discharge through CH_4, CO_2, and N_2 over a water surface. These structures demonstrate autocatalytic energy-dependent assembly in a strikingly biological manner. They closely resemble Archean microfossils and organized elements of carbonaceous chondrites not only in morphology, but in their chemically intractable kerogenous nature.

Simpler particulates have been studied before. Folsome and Morowitz (1969) showed the formation of membranous material by uv irradiation of alkanes upon aqueous solutions of phosphate and magnesium salts. Fox (1965, 1973) and Young (1965) have formed proteinoid microspheres assembled from thermally synthesized peptides, etc. Why have the complex structures reported here not been observed by others?

Apparently, most spark-discharge reactions are conducted either by semicorona or by arc-to-arc discharge. If present, water vapor, but not liquid water, is at the site of the discharge. Products formed in these reactions may condense in a distant water phase or upon the cool flask surface near the electrodes. In the latter case, a yellow-brown polymeric material is commonly seen. In our experiment, the spark is discharged directly upon a liquid water surface. Within hours, a yellow-brown scum appears near the single electrode upon the surface of the water. Examination of this early-formed material reveals organic microspheres. Partially hydrophobic polymeric material seems to be deposited first as membranous sheets upon the water surface. Motion of this surface, as by stirring, convection, or electrode movement, disturbs these sheets, which might then rearrange to spheroidal or other structured forms of greater stability.

Although tubular and sheet-like masses of material are seen, three groups of spheroids predominate. The largest consist of a complex gossamer mesh of membranous material. The smallest spheroids are composed of particulate subunits of 0.028-μm diameter. The intermediate spheroids show a complex structure containing both membranous and particulate matter. Two structural precursors are sufficient to explain the structures of the three groups of microspheres: membrane-like and particulate precursors. Interaction of both precursors, then, can account for the intermediate complex microspheres, whereas large and small microspheres represent predominantly either membranous or particulate precursors, respectively.

Exponential kinetics of formation of microspheres implies that the surface of one structure acts to initiate self-assembly of a second structure from similar precursors. Experiments are in progress to dissociate organic microspheres and to attempt specific reassembly.

It might seem surprising that organic microspheres have no traces of ATPase or H_2O_2 decomposing activities. Certainly for a protobiont one might expect some "haze" of potential catalytic activity. This lack could be due to inadequacy of our model system or to insensitivity of our enzyme assays. A more likely alternative is discussed (somewhat teleologically) below. An early protobiont could be synthesized with ease from the large array of abiotically available monomers and a great excess of thermal, uv, and electrical energy. One could argue that under these conditions a degradative metabolism would be of less use than a synthetic metabolism. In the early protobiont stage, selection would favor those structures with the greater rates of self-assembly. Thus, synthetic, rather than degradative, capabilities might be enhanced. Our laboratory is currently investigating the possibility that organic microspheres might catalyze such reactions.

Structures similar to synthetic organic microspheres might have played a necessary and significant role in the origin of life. They provide microbounded systems of complex internal morphologies and possess a biological range of high surface to volume ratios. They are synthesized with ease and are certainly the dominant repository of the input materials. Certainly, organic microspheres are not living, but they do have some fundamental biological attributes. Perhaps some of the early Archean microfossils could have been structures such as these.

Acknowledgments

This work was supported by a National Aeronautics and Space Admistration research grant, NGR-12-001-109, and by the University of Hawaii Research Council through the Biomedical Sciences Support Program. I thank Mrs. Marion Reed for assistance with the optical photomicrographs and Mr. Scott Kellogg and Mr. R. Paternek for their work on catalytic properties of organic microspheres.

REFERENCES

Barghoorn, E. S., and Schopf, J. W. 1966. *Science*, *152*:758.

Brooks, J., Muir, M. D.,and Shaw, G. 1973. *Nature(London)*, *224*;215.

Folsome, C. E., and Morowitz, H. J. 1969. *Space Life Sci.*, *1*:538.

Folsome, C. E., Allen, R. D., and Ichinose, N. K. 1975. *Precambrian Res.*, *2*:263.

Fox, S. W. 1965. In S. W. Fox (ed.), *The Origins of Prebiological Systems*, pp. 361 - 382. Academic Press, New York.

Fox, S. W. 1973. *Naturwissenschaften*, *60*:359.

Fraser, C. L., and Folsome, C. E. 1975. *Origins of Life*, in press.

MacGregor, A. M. 1941. *Trans. Geol. Soc. S. Afr.*, *43*:9.

Nagy, B., Zumberge, J. E., and Nagy, L. A. 1975. *Proc. Natl. Acad. Sci. USA*, *72*:1206.

Schopf, J. W., Oehler, D. Z., Horodyski, R. J., and Kvenvolden, K. A. 1971. *J. Paleontol.*, *45*: 477.

Schopf, J. W. 1975. *Ann. Rev. Earth Planet. Sci.*, *3*:213.

Studier, M. H., Hayatsu, R., and Anders, E. 1968. *Geochim. Cosmochim. Acta*, *32*:151.

Young, R. S. 1965. In S. W. Fox (ed.), *The Origins of Prebiological Systems*, pp. 347 - 357. Academic Press, New York.

18

COMPARISON OF LABORATORY SILICIFIED BLUE-GREEN ALGAE WITH PRECAMBRIAN MICROORGANISMS

S. FRANCIS, L. MARGULIS, AND W. CALDWELL
Boston University
ELSO S. BARGHOORN,
Harvard University

We have adapted a technique for the artificial silicification of wood (Barghoorn and Leo, 1974; Leo and Barghoorn, 1976) for the silicification of cyanophyte cultures. Artificial silicification of *lyngbya* has also been described, using a Ludox silica gel technique, by Oehler and Schopf (1971) and Oehler (1976). However, their system involved encasing algae at elevated temperatures (100°C) and pressures (3000 bar), whereas ours used only moderate conditions (room temperature to 55°C and 1 atm of pressure). We silicified a few grams each of the following cyanophytes: *Anabaena flos-aquae, Anabaena oscillatoria, Chloroglea microcystoides, Lyngbya* sp., *Chrococcidiopsis* sp., *Fischerella musicola, Oscillatoria formosa, Oscillatoria tenuis,* and *Cylindrospermum* sp. Monoalgal cultures of these were separated from the culture media, rinsed in distilled water, and immersed in tetraethylorthosilicate (Eastman Kodak Chemical Co.). These were then placed in closed containers in an oven at 55°C for between 3 to 5 days. The gelatinous material obtained was air dried. In the case of filamentous cultures, the solid material which resulted was coherent and could be easily fragmented, whereas coccoid cultures yielded material which, while hard, was composed of fine granules. The former was embedded in Lakeside Cement and ground with Carborundum until cells could be observed by light microscopy. The coccoid material, because of its granular nature, could not be treated in this way. It was placed in Permount, in which it could be satisfactorily observed without grinding. The variability observed in the quality of silicified material obtained may be attributed to these factors: (a) the amount of sheath material present; (b) the condition of the culture (in the case of *Chrococcidiopsis* sp., silicification was most successful using moribund cultures); (c) the volume of water present; (d) the

volume of tetraethylorthosilicate present; and (e) the temperature and length of time incubated. Silicification will occur
at room temperature, but months are required to produce the
same result as is produced in 1 week at 55°C.

A variety of "eucaryotic" characteristics has been simulated
here, including nearly "tetrahedral tetrads," trilete scars,"
and "nuclei" or other organellar inclusions. These will be
illustrated in the full paper (Francis *et al.*, 1976). We have
had thus far only limited success silicifying and observing
coccoid cyanophytes; however, coccoid characteristics were obtained by using a filamentous blue-green alga, *F. musicola*. In
young, actively growing cultures, *F. musicola* is a filament
approximately 10 m in diameter, consisting of subquadrate cells
surrounded by sheath. We have observed that over time these
filaments may break down to form single spherical cells or,
less frequently, diads, triads, and tetrads, also nearly spherical Occasionally the majority of filaments has been observed
to break down in this way to yield cultures almost indistinguishable from those of coccoid cyanophytes. The nearly tetrahedral configuration of cells observed in silicified *Fischerella*
is thought to have originated from displacement of four linearly
arranged spherical cells contained within a single sheath. The
nonmeiotic psudotetrad observed here appears to be remarkably
similar to *Eotetrahedron princeps* (Schopf and Blacic, 1971).
The pseudotrilete scar also observed in silicified material of
F. musicola is thought to be a desiccation artifact. Silicified
Lyngbya sp., *F. musicola*, *Chrococcidiopsis* sp., and *A. oscillatoria* have revealed evidence of shrunken cytoplasm and/or preserved granular material, features which were observed in live
cells before silicification and which may be mistaken as evidence
of eucaryotic organelles (Knoll and Barghoorn, 1975). Silicified
Lyngbya sheathes, sometimes of irregular diameter, have also
been observed.

The results of these experiments indicate that many of the
characteristics ascribed to eucaryotic microfossils may be simulated by using these procaryotic organisms. This work is intended to caution researchers against overinterpretation of
the Precambrian fossil record.

REFERENCES

Barghoorn, E. S. and Leo, R. F.: 1974.   Laboratory silifica-
    tion of plant tissues at low temperatures (abstr.). *Amer.
    J. Botany*, *61(s)* suppl. 13, June 16 - 20.
Francis, S., Margulis, L., Caldwell, W., and Barghoorn, E. S.
    1976.   Laboratory silicified blue-green algae compared with
    Precambrian microfossils, in preparation.
Knoll, A. H., and Barghoorn, E. S. 1975. *Science, 190:*52.
Leo, R. F., and Barghoorn, E. S. 1976. Silicification of

plant tissues under laboratory conditions, with special
reference to wood. Botanical Museum Leaflets, Vol. 25.
Harvard Univ. Press, Cambridge, Mass., in press.

Oehler, J. H., and Schopf, J. W. 1971. *Science, 174*:1229.

Schopf, J. W., and Blacic, J. M. 1971. *J. Paleontol. 45*;925 -
960.

19

EOASTRION AND THE METALLOGENIUM PROBLEM

ELSO S. BARGHOORN
Harvard University

Barghoorn and Tyler (1965) described the genus *Eoastrion* as a common member of the microbiota from the Lower Gunflint Iron Formation. Two species were distinguished based on minor differences of a common but distinctive morphology featured by an actinomorphic, asteriform structure three-dimensionally preserved. The organism was often present in profusion, most commonly three-dimensionally distributed in the hyaline chert matrix. The possible biological affinity of *Eoastrion* to a living counterpart did not become apparent until the dissemination of the work of Kuznetsov *et al.* (1963) and the translation of the work of Perfil'ev and Gabe (1969) describing the manganese- and iron-oxidizing organism *Metallogenium*, to which remarkable morphological comparison is evident (as subsequently noted by Cloud, 1965). More recent work on the budding bacteria (Hirsch, 1974) brings the putative phylogenetic relationship of *Eoastrion* and *Metallogenium* into closer focus and stimulates speculation as to the significance of *Eoastrion* to interpretation of the environment of the Middle Precambrian Gunflint sea. New occurrences of this organism from the Paradise Creek Formation (Kline, 1975) and its very recent discovery in Middle Precambrian rocks of the Duck Creek Dolomite of the Wyloo Group in northwestern Australia are discussed. Exceptionally well-preserved material in which the organism is extremely abundant is from the Upper Gunflint chert from a locality discovered by the author in 1968, but which has not been previously described. Optical sections through the chert, which is exceptionally transparent, show an average frequency of $7500/mm^3$ of discrete *Metallogenium* cells. The Upper Gunflint chert facies is featured by abundant carbonate crystals and slice pseudomorphs after halite, indicating its vigorous development in a hypersaline aqueous environment.

REFERENCES

Barghoorn, E. S., and Tyler, S. A. 1965. *Science, 147*:563.
Cloud, P. E. 1965. *Science, 148*:27.
Hirsch, P. 1974. *Ann. Rev. Microbiol., 28*:391.
Kline, G. 1975. *Geol. Soc. Amer. Cordilleran Sect.* (abstr.),March.
Kuznetzov, S. I., Ivanov, M. V., and Lyalikova, N. N. 1963. *Introduction of Geological Microbiology.* McGraw-Hill, New York.
Perfil'ev, B. V., and Gabe, D. R. 1969. *Capillary Methods of Investigating Microorganisms.* University of Toronto Press, Toronto

20

EVOLUTION OF MITOSIS AND THE LATE APPEARANCE OF METAZOA, METAPHYTA, AND FUNGI

L. MARGULIS
Boston University

The transition to the oxidizing atmosphere and the consequent ozone absorption of potentially lethal ultraviolet light are thought to be events that occurred in the Middle Precambrian (about 2×10^9 years ago; Cloud, 1974). It is argued that this atmospheric transition preceded by tens of millions of years the origin of eucaryotes and per se had nothing directly to do with the dramatic emergence and subsequent colonization of the land by metazoan and metaphytan organisms. Changing ultraviolet light and oxygen levels are significant environmental variables for the procaryotic organisms (blue-green algae=cyanobacteria, and bacteria) but not for most obligately aerobic eucaryotes. In fact, many natural protective mechanisms against uv light are known for microorganisms (Rambler et al., 1976 and this volume). Thus, the oxygenic atmosphere is thought to have been a necessary, but not sufficient, precondition for the origin of larger multicellular organisms. The evolution of regularized meiosis, on the other hand, a process which probably emerged in the Late Precambrian, is considered both necessary and sufficient to account for the late appearance of advanced tissue development characteristic of "higher" eucaryotes.

Although the interpretation of the Precambrian fossil record is fraught with difficulties, there is little doubt that Metazoa had evolved by approximately 0.7×10^9 years ago. Indisputable direct fossil evidence for eucaryotes prior to that date is not available (Knoll and Barghoorn, 1975; but also see Schopf, 1975). However, on biological grounds alone

(e.g., recognition of the eucaryotic nature of all animal and
green plant cells, comparison of ribosomal proteins, nucleo-
tide base--pair sequence studies of 5S RNA, and so forth;
Phillips and Carr, 1976), it is highly likely that the origin
of eucaryotic cells preceded by millions of years the appearance
of Late Precambrian soft-bodied fossil metazoans (Schopf, 1975).
Recent studies on the variations in mitosis and meiosis in
"primitive eucaryotes" suggest that these processes evolved
in protist microorganisms (multiauthored, *BioSystems*, Vol. 7,
No. 3, November, 1975). The variations in the mitotic systems
of amoebae, dinoflagellates, flagellate algae, radiolarians,
ciliates, amoeboflagellates, and others suggest that stabilization of the mitotic - meiotic system was a prerequisite for the
impressive morphological diversification of Metazoa and Metaphyta. This group of organisms, the protists, is set apart
from animals, plants, and fungi *(sensu stricto)* on the basis
of their significant mitotic and meiotic variations (Margulis,
1974, 1976). Furthermore, since mitotic microtubule systems
require low and regulated calcium ion concentrations (Borisy
et al., 1975), it is possible that the removal of calcium and
its external deposition evolved primarily for the intracellular
stabilization of microtubule-based morphogenetic systems, including that of the mitotic apparatus, and thus preadapted
many organisms for the formation of protective external calcium
carbonate hard parts. This idea is supported by the observation
that microtubules of the axopods of multichambered foraminifer-
rans are involved in the calcium carbonate deposition of the
shells (Grell, 1973).

     In summary, the following sequence of events is suggested:
(a) the origin of oxygen-eliminating photosynthetic metabolism
in cyanobacteria; (b) the transition to the oxygenic atmosphere
and the formation of the ozone layer; (c) the convergent origin
of oxygen-utilizing metabolic pathways in many microbial groups;
(d) the origin of microtubule-based morphogenetic systems in
early protists; (e) the convergent evolution of mitotic and
meiotic genome distribution mechanisms in protists; (f) the
active metabolic removal of calcium from the sites of microtubule assembly and the deposition of calcium carbonate ex-
ternally in several groups of organisms; (g) the stabilization
of "classical" mitotic - meiotic patterns in several groups
of protists ancestral to metazoans and metaphytans; and finally
(h) the adaptive radiation of early metazoans, including those
with fossilizable calcium carbonate hard parts.

REFERENCES

Borisy, G. G., Marcum, J. M., Olmsted, J. B., Murphy, D. B.,
     and Johnson, K. A. 1975. *NY Acad. Sci.*, *253*:107.

Cloud, P. E., Jr. 1974. *Amer. Sci., 62*:54.

Grell, K. 1973. *Protozoology.* Springer - Verlag, Heidelberg and New York.

Knoll, A., and Barghoorn, E. S. 1975. *Science, 190*:52.

Margulis, L. 1974. In *Handbook of Genetics* Vol. 1, pp. 1 - 41. Plenum, New York.

Margulis, L. 1976. *Taxon, 25*:391.

Phillips, D. O., and Carr, N. C. 1976. *Taxon,* in press.

Rambler, M., Margulis, L., and Walker, J. C. G. 1976. A reassessment of the roles of oxygen and ultraviolet light in Precambrian evolution   *Nature* (in press)

Schopf, J. W. 1975. *Ann Rev. Earth Planet. Sci. 3*:213.

21

IRON–SULFUR PROTEINS AND SUPEROXIDE DISMUTASES IN THE EVOLUTION OF PHOTOSYNTHETIC BACTERIA AND ALGAE

D.O. HALL, J. LUMSDEN, AND E. TEL-OR
University of London King's College

The early evolution of life probably involved the development of bacteria from the obligate anaerobic-fermenting types through the photosynthetic bacteria to the blue-green algae and subsequently to the green algal types. The question arises as to whether it is at all possible from Precambrian microfossil evidence to decide whether there is a logical sequence of development from bacteria (of whatever type) to algae (also of whatever type). It will probably be very difficult from the morphology of microfossils of Precambrian rocks to decide whether different types of bacteria preceded others, since there is very little distinguishing morphological evidence between the different bacteria thought to be involved, unless the filamentous photosynthetic bacteria and blue-green algae can be readily distinguished. However, at this stage the morphologies are uncertain and are very difficult to interpret from microfossil evidence (Muir and Hall, 1974). This is not to say that microfossils will not be found in Precambrian rocks which would provide the evidence, but it is missing at present (see the review by Schopf, 1975).

It seems as if at the present stage the biochemical evidence from specific proteins within bacteria and algae may provide the best evidence for deriving some logical development from the anaerobic bacteria to the aerobic algae. A combination of biochemical and morphological evidence will probably ultimately be the most convincing evidence for this stage of early evolution in the Precambrian (Broda, 1975; Hall *et al.*, 1975b). However, in this paper we are going to discuss evidence based solely on the properties of two enzymes, namely the Fe - S proteins (the best known example of which are ferredoxins) and the superoxide dismutases (enzymes which protect organisms against the interaction with toxic free radical superoxide). In the case of the Fe - S proteins, these are the only proteins that have thus far been sequenced which are known to occur in all living organisms. With the superoxide dismutases (SOD) we have a unique advantage of having at least two different types of this enzyme [cyanide sensitive (Cu/Zn) and cyanide insensitive (Fe or Mn)] which are distinct in procaryotic and eucaryotic microorganisms. Both of these enzymes thus offer interesting possibilities for studying early evolution across the anaerobic/aerobic and procaryotic/eucaryotic transitions.

I.   THE IRON-SULFUR PROTEINS

All Fe - S proteins (Hall *et al.*, 1975b; Holm, 1975; Lovenberg, 1973) contain nonhaem iron atoms bonded to sulfur atoms --both cysteine sulfurs of the protein and inorganic sulfide atoms (except in rubredoxin). The most common types are the ferredoxins, which in all cases contain equivalent amounts of iron and sulfur in their active centers with the iron bonded to cysteine sulfurs (see Fig. 1 for the structures of the Fe - S centers of ferredoxins and the one-iron center of rubredoxin). Rubredoxins occur in bacteria, but as yet not as much is known about their properties and role, so that we will not discuss them in the rest of this article. Hopefully, this lack of information will be rectified in the near future, as one-iron rubredoxins are undoubtedly of physiological importance in bacterial metabolism.

Ferredoxins have been proposed to be among the earliest proteins formed, since they have the following properties:

(1) They transfer electrons at very low redox potentials, close to that of the hydrogen electrode (see Table 1 for various properties).

(2) They catalyze reactions of physiological importance in obligate anaerobic bacteria such as hydrogen uptake and evolution, ATP formation, pyruvate metabolism, nitrogen fixation, and photosynthetic electron transport (Table 2).

(3) They are small proteins consisting of relatively few stable and simple amino acids, which can be easily formed under abiogenic primitive earth conditions. They contain no (or small amounts of) histidine, methionine, tryptophan, and tyrosine.

(4) The active center consists of only iron and sulfur and the holoprotein can be reconstituted *in vitro* from the apoprotein using inorganic iron and sulfur under nonenzymatic, anaerobic conditions. The 4Fe - 4S or 2Fe - 2S active center which is formed can also be extracted intact from the protein itself and kept in solution and also subsequently added back to the

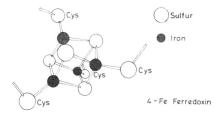

Sulfur

Iron

4 - Fe Ferredoxin

2 - Fe Ferredoxin

Rubredoxin

Fig. 1 Structures of the 4Fe - 4S centers and 2Fe - 2S centers in ferredoxins, and the 1Fe center in rubredoxin (Hall et al., 1975b, Holm, 1975).

protein.

(5) Ferredoxins catalyze electron-transfer reactions from a potential of -600 to +400 mV. This ability of the ferredoxins to catalyze such diverse redox-potential reactions can be manipulated in the laboratory by altering the protein configuration around the active center and has led to the proposal of the three-state hypothesis for the 4Fe - 4S active center (Fig. 2).

(6) The positions of the cysteines, which hold the iron in a specific environment, are invariant in the different classes of ferredoxins; this is very useful in sequence work and the analysis of evolution based on such sequence data (Fig. 3).

(7) The Fe - S proteins can occur both in soluble and membrane-bound forms and can catalyze electron-transfer reactions under both conditions.

(8) Fe - S proteins can form complex electron-transfer proteins with flavins and molybdenum (and possibly also haems), e.g., nitrogenase and xanthine dehydrogenase.

The sequence of amino acids in ferredoxins has given us quite some insight into the evolution of ferredoxins. Figure 3 shows the sequences of representative ferredoxins containing 8Fe, 4Fe, and 2Fe per molecule. It will be seen that the 8Fe ferredoxins contain two homologous sections of about 28 amino acids each, each section of which contains four invariant cys-

TABLE 1
Properties of Representative Iron – Sulfur Proteins[a]

	Active group (atoms per molecule)	Molecular weight	Number of Amino Acids	Redox potential (E_0' in mV)	Number of electrons transferred
4Fe + 4S center					
8Fe ferredoxins[b]					
Clostridium (obligate anaerobic bacterium)	8Fe, 8S	6,000	55	-395	2
Chlorobium (green photosynthetic bacterium)	8Fe, 8S	6,000	60	---	---
Chromatium (red sulfur photosynthetic bacterium)	8Fe, 8S	10,000	81	-490	2
Rhodospirillum rubrum I (red nonsulfur photosynthetic bacterium)	8Fe, 8S	13,000	---	---	---
Azotobacter III (aerobic N_2-fixing bacterium)	8Fe, 8S	15,000	130	-420	---
4Fe ferredoxins[c]					
Desulphovibrio (anaerobic SO_4-reducing bacterium)	4Fe, 4S	6,000	56	-330	1
Bacillus (facultative N_2-fixing bacterium)	4Fe, 4S	8,000	78	-380	1
4Fe HiPIP					
Chromatium	4Fe, 4S	9,650	86	+350	1
2Fe + 2S center					
2Fe ferredoxins[d]					
Spinach (higher plant)	2Fe, 2S	10,600	97	-420	1
Microcystis (blue-green alga)	2Fe, 2S	10,300	98	---	1
Scenedesmus (green alga)	2Fe, 2S	10,600	96	---	1
Azotobacter I (aerobic N_2-fixing bacterium)	2Fe, 2S	21,000	181	-350	1
Pseudomonas putida (aerobic bacterium)	2Fe, 2S	12,500	114	-240	1
E. coli (aerobic bacterium)	2Fe, 2S	12,600	---	-360	--
Pig adrenals (mammal)	2Fe, 2S	12,500	115	-270	1
Mitochondria, Complex III (mammalian)	2Fe, 2S	30,000	---	+280	1

1Fe center					
1Fe rubredoxin[e]					
Clostridium	1Fe	6,000	54	-60	1
Complex Fe - S proteins					
Mitochondrial succinate dehydrogenase (mammalian)	4Fe, 4S, 1FAD	70,000 (dimer) --	--	--	--
Mitochondrial NADH dehydrogenase (mammalian)	28Fe, 28S, 1FMN	--	--	--	--
Xanthine oxidase (milk, bacteria)	8Fe, 8S, 2FAD, 2 Mo	275,000	--	--	--
Nitrogenase (Clostridium or Klebsiella):					
Molybdenum iron protein	18 or 24 S, 2 Mo	220,000	--	-60 and -280 mV	--

[a] For original references see Hall et al., 1975c.

[b] Ferredoxins with 8FE + 8S have also been reported in: C. pasteurianum, C. acidi urici, C. butyricum, C. tartarivorum, C. tetanomorphum, C. thermosaccharolyticum, Peptococcus aerogenes, Peptostreptococcus elsdenii, and Veillonella alcalescens.

[c] Ferredoxins with 4Fe - 4S have also been reported in: B. polymyxa, B. stearothermophilus, Spirochaeta aurentia, spinach chloroplast membrane, D. gigas, D. desulfuricans, and Rhodospirillum rubrum. High-potential iron - sulfur protein has been reported in Rhodopseudomonas.

[d] Ferredoxins with 2Fe + 2S have also been reported in: Aethusa, Agrobacterium, Amaranthus, Anabaena, Anacystis, Aphanothece, Brassica, Botrydiopsis, Bumilleriopsis, Chenopodium, Chlamydomonas, Chlorella, Cladophora, Clostridium (azoferredoxin and E.p.r. protein), Colocasia, Cyanidium, Cyperus, Datura, Equisetum, Euglena, Gossypium, Laminum, Leucaena, Medicago, Navicula, Nostoc, Petroselenium, Phaseolus, Phormidium, Pinus, Pisum, Polystichum, Porphyridium, Prophyrathenera, Rhizobium, Sambucus, Spirulina, Stelluria, Tolypothrix, Zea, pig adrenals, and pig testes.

[e] Rubredoxins have also been reported in: Chloropseudomonas ethylica, Clostridium butyricum, Clostridium stricklandii, Desulfovibrio desulfuricans, Desulfovibrio gigas, Peptococcus aerogenes, Peptococcus glycinophilus, Peptostreptococcus elsdenii, and Pseudomonas oleovorans.

TABLE 2

Some Electron-Transfer Reactions Involving Fe – S Protein[a]

Type of reaction	Representative organism	Type of Fe – S protein
1. Phosphoroclastic reactions: Pyruvate + P_i $\xrightarrow{CoA}$ acetyl phosphate + CO_2	*Clostridium*	8Fe
2. Synthesis of α–keto acids (CO_2 fixation), e.g.: Acetyl–CoA + CO_2 → pyruvate + CoA Succinyl–CoA + CO_2 → α–oxoglutarate + CoA Propionyl–CoA + CO_2 → α–oxobutyrate + CoA	Fermentative and photosynthetic bacteria, e.g., *Clostridium, Chromatium*	8Fe
3. One-carbon metabolism: $CO_2 \rightleftarrows HCO_3^-$	*Clostridium*	8Fe
4. Hydrogen metabolism: $2H^+ + 2e^-$ $\xrightarrow{Hydrogenase}$ H_2	Certain algae; certain bacteria	12Fe
5. Nitrogen fixation: $N_2 + 3H_2 \rightleftarrows 2NH_3$	Fermentative and photosynthetic bacteria	(a) 18 – 24Fe/mol (b) 2Fe/mol
6. Nicotinamide nucleotide oxidoreduction: NADH + $NADP^+ \rightleftarrows NAD^+$ + NADPH	Anaerobic bacteria	8Fe
7. Photosynthetic electron transfer in bacteria	*Chromatium*	*4Fe* (Ferredoxin + HiPIP)
8. α–Hydroxylation of hydrocarbons: $PCH_3 + NADH + H^+ + O_2 \rightarrow RCH_2OH + NAD^+ + H_2O$	*Rhodopseudomonas* *Ps. oleovorans*	*1Fe* (Rubredoxin)
9. Sulfite reduction: $SO_3^{2-} \rightarrow S^{2-}$	Algae, plants *D. gigas, E. coli,* *C. pasteurianum*	
10. Nitrite reduction: $NO_2^- \rightarrow NH_3$	Algae, plants, *Micrococcus* *denitrificans*	

196

TABLE 2 (continued)

11.	Nicotinamide nucleotide reduction: $NADP^+ + H_2O \rightarrow NADPH + \frac{1}{2}O_2$ Photophosphorylation: $ADP + P_i \rightarrow ATP$	Plants and algae	2Fe 2Fe
12.	(a) Oxidation of $NADH_2$ by mitochondria: $NADH_2 + \frac{1}{2}O_2 \rightarrow NAD + H_2O$ ATP		
	(b) Oxidation of succinate: Succinate $\rightarrow$ fumarate ATP	Plants; mammals; bacteria	Two Fe − S subunits
13.	Oxidation of xanthine and aldehydes: $R{-}H + H_2O + O_2 \rightarrow R{-}OH + H_2O_2$	Bacteria; mammals	8Fe/mol
14.	Hydroxylation: $R{-}H + O_2 + NAD(P)H \rightarrow ROH + H_2O + NAD(P)^+$	Mammalian (adrenal mitochondria)	2Fe

[a]For original references, see Hall *et al.*, 1975c.

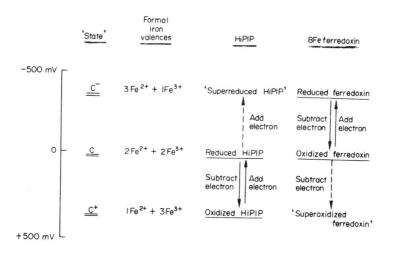

Fig. 2 *The "three-state hypothesis" for the redox poten-*
tials of electron transfer in the 4Fe - 4S active centers of
ferredoxins and HiPIP. Note the difference in redox potential
over a 1 V range (see Hall et al., 1975b for a detailed dis-
cussion).

teines. There seems to be a logical progression from the Clostridial-type ferredoxin through the green photosynthetic bacterial ferredoxin (*Chromatium*), to the 4Fe ferredoxin from the
sulfate reducer *Desulfovibrio*. The 4Fe ferredoxins have only
been sequenced thus far in the sulfate reducing bacteria which
are thought to occur at the transition between anaerobic and
aerobic environments. They have also been recently sequenced in
B. stearothermophilus (only 4 cysteines present). It has been proposed that the first half of the *Desulfovibrio* ferredoxin is
analogous to the first half of the 8Fe ferredoxins, while the
second half of the *Desulfovibrio* ferredoxin may be somewhat similar to that of the 2Fe ferredoxins from algae and plants.
      The 2Fe ferredoxins which have been sequenced so far all
contain four invariant cysteines. The cysteine at position 18
is missing from the blue-green alga *Aphonothece* and the primitive plant *Equisetum*. The question of the transition from the
sulfate-reducing bacteria (4Fe ferredoxin) to the ferredoxin
(2Fe) from the blue-green algae is still an enigma, but hopefully we will find organisms that span this transition, which
may indicate whether 4Fe or 2Fe ferredoxins were the important
ferredoxins in the anaerobic environment prevalent during the
development of organisms up to the sulfate reducers. If these
organisms are found, they may also indicate which ferredoxins
were prevalent in the aerobic organisms present after the development of the oxygen-evolving photosynthesis of the blue-

FIG. 3. Amino-acid sequences of representative ferredoxins.

```
        1                                                                    28
A  Ala-Phe-Val-Ile-    Asn-Asp-Ser-Cys-Val-Ser-Cys-Ala-Gly-Ala-Cys-Ala-Gly-Glu-Cys-Pro-Val-Ser-Ala-Ile-Thr-Gln-Gly-Asp-Thr-
B  Ala-Leu-Tyr-Ile-    Thr-Glu-Glu-Cys-Thr-Tyr-Cys-Gly-Ala-Cys-Glu-Pro-Glu-Cys-Pro-Val-Thr-Ala-Ile-Ser-Ala-Gly-Asp-Asp-
C  Ala-Leu-Met-Ile-    Thr-Asp-Gln-Cys-Ile-Asn-Cys-Asn-Val-Cys-Gln-Pro-Glu-Cys-Pro-Asn-Gly-Ala-Ile-Ser-Gln-Gly-Asp-Glu-
D  ----- Pro-Ile-Gln-  Val-Asp-Asn-Cys-Met-Ala-Cys-Ile-Asn-Glu-Cys-Ile-Asn-Glu-Cys-Pro-Val-Asp-Val-Phe-Gln-Met-Asp-Glu-Gln-

        29                                                            55
A  Gln-Phe-Val-Ile-Asp-Ala-Asp-Thr-Cys-Ile-Asp-Cys-Gly-Asn-Cys-Ala-Asn-Val-Cys-Pro-Val-Gly-Ala-Pro-Asn-Gln-Glu
B  Ile-Tyr-Val-Ile-Asp-Ala-Asn-Thr-Cys-Asn-Glu-Cys-Val-Ala-Val-Cys-Pro-Ala-Glu-Cys-Ile-Val-Gln-Gly (60)
                                                   Gly-- Leu-Asp      Glu-Gln
                                                       30
C  Thr-Tyr-Ile-Glu-Pro-Ser-Leu-Cys-Thr-Glu-Cys-Val-Gly-Val-Asp-Cys-Val-Gln-Val-Cys-Pro-Ile-Lys-Asp-Pro-Ser-His-Glu----Gly (81)
                                                                Val-Cys
                        Gly-His-Tyr-Glu-Thr-Ser-Glu
D  Gly-Asp-Lys-Ala-Val-Asn-Ile-Pro-Asn-Ser-Asn-Leu-Asp-Asp-Glu-Cys-Val-Glu-Ala-Ile-Gln-Ser-Cys-Pro-Ala-Ala-Ile-Arg-Ser (56)

        1                                                                    30
E  Ala-Ser-Tyr-Lys-Val-Thr-Leu-Lys-Thr-Pro-Asp-Gly-Asp-Asp-Val-Ile-Thr-Val-Pro-Asp-Asp-Glu-Tyr-Ile-Leu-Asp-Val-Ala-Glu-Glu-
F  Ala-Thr-Tyr-Lys-Val-Thr-Leu-Lys-Thr-Pro-Ser-Gly-Asp-Gln-Thr-Ile-Glu-Cys-Pro-Asp-Asp-Thr-Tyr-Ile-Leu-Asp-Ala-Ala-Glu-Glu-
G  ----- Tyr-Lys-Val-Thr-Leu-Val-Thr-Pro-Ser-Gly-Thr-Gln-Val-Glu-Phe-Thr-Leu-Asp-Val-Pro-Glu-
H  Ala-Ser-Tyr-Lys-Val-Lys-Leu-Val-Thr-Pro-Glu-Gly-Thr-Gln-Glu-Phe-Glu-Cys-Pro-Asp-Asp-Val-Tyr-Ile-Leu-Asp-His-Ala-Glu-Glu-

        31                                              50               60
E  Glu-Gly-Leu-Asp-Leu-Pro-Tyr-Ser-Cys-Arg-Ala-Gly-Ala-Cys-Ser-Thr-Cys-Ala-Gly-Lys-
F  Ala-Gly-Leu-Asp-Leu-Pro-Tyr-Ser-Cys-Arg-Ala-Gly-Ala-Cys-Ser-Ser-Cys-Ala-Gly-Lys-
G                         Ser-Cys-Arg-Ala-Gly-Ala-Cys-Ser-Ser-Cys-Leu-Gly-Lys-
H  Glu-Gly-Ile-Val-Leu-Pro-Tyr-Ser-Cys-Arg-Ala-Gly-Ser-Cys-Ser-Ser-Cys-Ala-Gly-Lys-Val-Ala-Ala-Gly-Glu-Val-Asn-Gln-Ser-Asp-

        61                                                          90
H  Gly-Ser-Phe-Leu-Asp-Asp-Asp-Gln-Ile-Glu-Glu-Gly-Trp-Val-Leu-Thr-Cys-Val-Ala-Tyr-Ala-Lys-Ser-Asp-Val-Thr-Ile-Glu-Thr-His-
        91
H  Lys-Glu-Glu-Glu-Leu-Thr-Ala (97)
```

FIG. 3. Amino-acid sequences of representative ferredoxins. See Hall et al. (1975a) and Lovenberg (1973) for original references, and see the text. Key: A, Clostridium butyricum (8Fe), obligate anaerobic fermenting bacterium; B, Chlorobium limicola (8Fe), green photosynthetic bacterium; C, Chromatium (8Fe), purple (sulfur) photosynthetic bacterium; D, Desulphovibrio gigas (4Fe), sulfate-reducing bacterium; E, Aphanothece (2Fe), blue-green alga; F, Scenedesmus (2Fe), green alga; G, Equisetum (3Fe), primitive plant; H, Medicago (2Fe) alfalfa higher plant.

green algae. The occurrence of 2Fe ferredoxins in bacteria is known, for example, in *Clostridium* and in *Escherichia coli,* but as yet we don't have the sequence of any of them to throw any light on this problem.

In studying the comparative structural and metabolic characteristics of red photosynthetic bacteria and blue-green algae (Table 3) it is quite striking that one can find red photosynthetic bacteria which have properties similar to blue-green algae. Similarly, studying blue-green algae, we find algae similar to red photosynthetic bacteria at one end and blue-green algae similar to the eucaryotic green algae at the other end of the spectrum. Thus, it is possible that there is a logical transition from the photosynthetic bacteria to the blue-green algae, with the sulfate reducers falling somewhere in between (in parallel?) as the environment became somewhat aerobic with the (slow?) development of blue-green algal photosynthesis.

The technique of studying the interaction between specific antibodies of ferredoxins and different ferredoxins themselves provides an interesting insight into the specificity of ferredoxin properties and to the evolution of ferredoxins (Tel-Or *et al.,* 1975). Figure 4 shows some recent work which we have done on the interaction between five different ferredoxin antibodies (including 8Fe, 4Fe, and 2Fe ferredoxins) from bacteria, blue-green algae, green algae, and higher plants and their interaction with 28 different ferredoxins representative of bacteria (both photosynthetic and nonphotosynthetic), blue-green algae, red and green algae, primitive plants, and higher plants. It will be seen from the histograms that there is a selectivity between the antibodies of given ferredoxins and their ability to interact with different types of ferredoxins. This gives us a clue to the development and specificity of ferredoxins. Even though the state of this art is still rather "primitive," this technique is providing us with evidence of evolutionary development and specificity of ferredoxins all the way from primitive obligate anaerobic fermenters to higher plants and animals.

In Fig. 5 we show a possible evolutionary scheme of organisms based solely on the properties of ferredoxins. Even though this scheme may seem rather simpleminded, it does show a sequence which is based solely on the knowledge of a specific protein, and since this protein is the only ubiquitous protein so far examined in detail in both obligate anaerobes and photosynthetic organisms, it is a beginning which seems to provide us with some information on Precambrian evolution.

II. SUPEROXIDE DISMUTASES

Oxygen in the atmosphere is a by-product of the evolution of the O_2-evolving photosynthesis first developed by the blue-green

TABLE 3

Some Comparative Properties of Red Photosynthetic Bacteria and Blue-green Algae[a]

	Photosynthetic membranes	Growth conditions	Characteristics
Photosynthetic bacteria			
1. *Rhodospirillum rubrum*	Vesicles	Anaerobic, light	No difference in photosynthetic apparatus and bacteriochlorophyll content
Rhodopseudomonas viridis	Lamellae	Anaerobic, dark	Respiratory system present in both but terminal oxidase different; vesicles and bacteriochlorophyll only under anaerobic conditions
2. *Rhodospirillum rubrum*	Vesicles	Anaerobic, light	
		Aerobic, dark	
3. *Rhodospirillum rubrum*	Vesicles	Anaerobic, light	8Fe ferredoxin + 4Fe ferredoxin
		Aerobic, dark	4Fe ferredoxin only
4. *Rhodopseudomonas spheroides*	Vesicles	Anaerobic, light	Bacteriochlorophyll + cytochrome o
		Aerobic, dark	No bacteriochlorophyll or vesicles + cytochrome a.
5. *Rhodopseudomonas capsulata*		Anaerobic, dark	Bacteriochlorophyll, but fewer vesicles
		Anaerobic, light	Phosphorylation coupling factors interchangeable
6. *Ectothiorhodospira mobilis*	Stacked lamellae	Aerobic, dark	
		Anaerobic, light (high concn. of salt and temperature)	

TABLE 3 (continued)
Some Comparative Properties of Red Photosynthetic Bacteria and Blue-green Algae

	Photosynthetic membranes	Growth conditions	Characteristics
Blue-green algae			
1. Facultative photoheterotrophs		Grow in light with CO_2 or glucose as C source	
Facultative chemoheterotrophs		Grow in light or dark with glucose as C source	
2. Fatty-acid biosynthesis		Desaturating enzyme systems	
		(a) Only bacterial type	
		(b) Both bacterial and algal types	
		(c) Only algal type	
3. Polyunsaturated fatty-acid content		Species diversity	
		(a) Low content (similar to other procaryotes)	
		(b) Intermediate	
		(c) High content (chloroplast-like)	
4. Glycolipid content (mono- and digalactosyl diglyceride)		Species diversity	
		(a) Blue-green algae and chloroplasts contain both	
		(b) Green photosynthetic bacteria contain mono- form	
		(c) Red photosynthetic bacteria contain neither	
5. Dependence on NADP as electron carrier		Species diversity	
		(a) Pentose phosphate patyway for substrate oxidation	
		(b) Tricarboxylic acid cycle for biosynthesis	
		(c) NADPH oxidation yields ATP	
6. pH requirements		Species diversity	
		(a) Blue-green algae do not grow below pH 4-5	
		(b) Eucaryotic algae grow below pH 4 (below 56°C)	

[a] For original references, see Carr and Whitton, 1973 and Hall et al., 1975a.

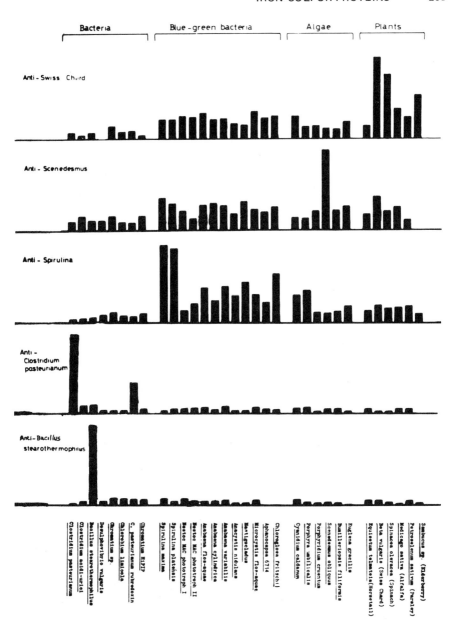

FIG. 4.   *Ferredoxin antibody interactions with 28 different ferredoxins (Tel-Or et al., 1976).*

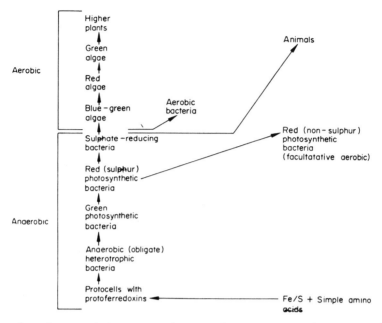

Fig. 5 Possible evolutionary development of ferredoxins (Hall et al., *1975b).*

algae. It is a prerequisite for the development of life as we know it, but it is not often realized that O_2 itself is, in fact, quite a toxic compound and that organisms have had to develop quite specific protection mechanisms against high concentrations of O_2. A key enzyme involved in this protection mechanism seems to be the recently recognized superoxide dismutase (SOD) (Fridovich, 1975; Halliwell, 1974; Lumsden, 1975). Even though nearly all of the oxygen in our earth's atmosphere is derived from O_2-evolving photosynthesis as a result of the splitting of water, there was undoubtedly O_2 present in small quantities during the early evolution of life as a result of the uv-induced photolysis of water (the so-called Urey effect). Thus, it is quite probable that even obligate anaerobic bacteria (nonphotosynthetic and photosynthetic) had to acquire some way of protecting themselves against the toxic effects of O_2 at a very early stage, before the development of O_2-evolving photosynthesis (Lumsden, 1975; Morris, 1975). The occurrence of SOD was originally thought to be confined to aerobic organisms, but discoveries since 1975 or so have shown that this enzyme also occurs in obligate anaerobic bacteria, e.g.,*Clostridium, Desulfovibrio, Chlorobium, Chromatium,* etc. (Hewitt and Morris, 1975; Lumsden and Hall, 1975b; Morris, 1975). This provides us with an interesting enzyme for studying the evolution of O_2 protection and O_2 evolution, both of which are key phenomena in the development of life.

The interaction of O_2 with free radicals produces superoxide (O_2^-). Figure 6 shows the reaction of O_2 with electrons
and the enzymes involved in breading down superoxide to peroxide
(O_2^{2-}) and subsequent reactions to hydroxyl radicals $(OH\cdot)$. The
toxicity of peroxide and superoxide is considered to be high,
with superoxide more so than peroxide. However, hydroxyl radicals are thought to be even more toxic. The enzymes SOD and
catalase undoubtedly had to be present to take care of the O_2
radicals produced from an early stage. It is interesting that
catalase is a haem enzyme which doesn't seem to occur in the obligate anaerobic fermenters like Clostridia, but SOD does occur
in these organisms, much to many biologists' surprise. The primary electron acceptor(s) in bacterial and plant-type photosynthesis seem to be able to interact with O_2 in bacterial and plant-
type photosynthesis to produce superoxide (Allen and Hall, 1974;
Boucher and Gringras, 1975).

There are at least three different types of SOD which we can
distinguish, and these are shown in Table 4. The Fe- and Mn-containing SODs are both cyanide insensitive and, until recently,
were thought to occur only in procaryotic organisms. The Cu + Zn
- containing SOD is cyanide sensitive, and, again until recently,
was only thought to occur in eucaryotes. This hypothesis was very
neat, but unfortunately it does not seem to be quite correct.
What has happened since 1975 or so is that algae of the eucaryotic type (reds, browns, and greens) have been shown (Asada et al.,
1975; Lumsden, 1975) to contain the Fe or Mn type of SOD and,
second, fungi and protozoa, which are eucaryotes, have been shown
(Lavelle et al., 1974; Lindmark and Müller, 1974) to contain the
cyanide-insensitive, Mn-containing SOD. Thus, the hypothesized
distinction between eucaryotes and procaryotes does not seem to
be universal. From our point of view this is in fact very for-
tunate, since we now have two distinct types of enzyme which span
the procaryote/eucaryote transition and also the anaerobic/aerobic
transition.

It has also been shown that two different SODs are present
in mitochondria (Fridovich, 1975). The mitohcondrial SODs are of
the cyanide-insensitive Fe and Mn type; the Mn SOD has an amino
acid sequence very similar to that from bacteria (see Fig. 7).
This provides interesting support for the theory of the symbiotic
origin of mitochondria from bacteria. A survey of SOD in photosynthetic bacteria and various algae (Fig. 8) shows that the cyanide-insensitive type of enzyme occurs in all photosynthetic
organisms from the red and green photosynthetic bacteria to the
green algae. The ability to isolate intact chloroplasts from the
coenocytic green alga Codium allowed us to show the localization
of a cyanide-insensitive SOD in the chloroplast itself (Lumsden
and Hall, 1975b). The higher plant chloroplasts do not contain
the same type of SOD as has so far been found in green and red

TABLE 4
Properties of Superoxide Dismutases[a]

Enzyme type	Metallion content (g-atom mole^{-1})	Molecular weight	Subunit structure
Cu – Zn			
Many eucaryotes, including mammalian tissues, yeast, Neurospora, and green plants	2Cu 2Zn	32,000	α_2
Photobacterium leiognathi	1Cu 2Zn	33,100	α_β
Mn			
Escherichia coli	1.4	39,500	α_2
Streptococcus mutans	1.2	40,250	α_2
Rhodopseudomonas spheroides	1.1	37,400	α_2
Bacillus stearothermophilus	––	40,000	α_2
Thermus aquaticus	2	80,000	α_4
chicken liver mitochondria	2.1	80,000	α_4
Yeast mitochondria	4	96,000	α_4
Pleurotus plearius (fungus)			
SODc	2.1	76,000	α_2 β_2
SODm	2	78,000	α_4
Fe			
Escherichia coli	1.0	38,700	α_2
Photobacterium leiognathi	1.6	40,600	α_2
Plectonema boryanum	0.9	36,500	α_2
Spirulina platensis	1.0	37,400	α_2

[a]For original references see Fridovich, 1975 and Lumsden, 1975.

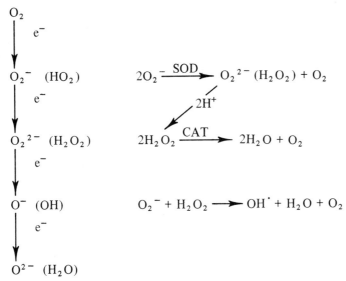

$$O_2 \xrightarrow{e^-} O_2^- \ (HO_2) \xrightarrow{e^-} O_2^{2-} \ (H_2O_2) \xrightarrow{e^-} O^- \ (OH) \xrightarrow{e^-} O^{2-} \ (H_2O)$$

$$2O_2 \xrightarrow{\text{SOD}} O_2^{2-} \ (H_2O_2) + O_2$$

$$\searrow 2H^+$$

$$2H_2O_2 \xrightarrow{\text{CAT}} 2H_2O + O_2$$

$$O_2^- + H_2O_2 \longrightarrow OH^{\cdot} + H_2O + O_2$$

Fig. 6. Oxygen, Superoxide and Catalase Reactions (SOD= superoxide dismutase; CAT= catalase). (Halliwell, 1976, Lumsden, 1975)

			5		10		15
B. stearothermophilus	Mn	Pro–Phe–Glu–Leu–Pro–Ala–Leu–Pro–Tyr–Pro–Tyr–Asp–Ala–Leu–Glu					
E. coli	Mn	Ser–Tyr–Thr	Ser		Ala		
E. coli	Fe	Ser			Ala Lys		Ala
Chicken liver mitochondria	Mn	Lys–His–Thr	Asp		Asp	Gly	
Bovine erythrocyte	Cu/Zn	Ac–Ala–Thr–Lys–Ala–Val–Cys–Val–Leu–Lys–Gly–Asp–Gly–Pro–Val–Gln					

			20		25		
B. stearothermophilus	Mn	Pro–His–Ile–Asp–Lys–Glu–Thr–Met–Asn–Ile–His–His–Thr–Lys					
E. coli	Mn	Phe	Gln		Glu–Leu ? ?		
E. coli	Fe	Ser–Ala	? Ile–Glu–Tyr		Tyr–Gly		
Chicken liver mitochondria	Mn	Ser–Ala	Ile		Gln–Leu ? ?		
Bovine erythrocyte	Cu/Zn	Gly–Thr–Ile–His–Phe–Glu–Ala–Lys–Gly–Asp–Thr–Val–Val–Val					

Fig. 7. Comparison of the N-terminal region of superoxide dismutase from B. stearothermophilus with the N-terminal regions of two other bacterial and one mitochondrial dismutase. Residues identical to those in the B. stearothermophilus sequence have not been included. The corresponding region of the bovine erythrocyte enzyme is also shown. (Budgen et al., 1975)

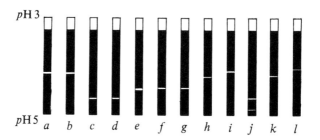

Fig. 8.   Polyacrylamide gel isoelectric focusing of SOD
(cyanide-insensitive) from photosynthet-c organisms.   From left
to right, the organisms are: a. Chlorobium thiosul hatophilum
(green bacterium); b. Chromatium D (purple, sulphur bacterium);
c. Rhodopseudomonas spheroides (purple, non-sulphur bacterium)
anaerobically grown; d, Rps. spheroides, aerobically grown; e,
Nostoc muscorum (blue-green alga); f, Anabaena cylindrica (blue-
green alga); g, A. flos-aquae (blue-green alga); h, Spirulina
platensis (blue-green alga); i, Porphyridium cruentum (red alga);
j, Scenedesmus obliquus (green alga); k. Codium fragile (coenocytic green alga) chloroplast stromal extract; l, C. fragile,
cytoplasmic extract. (Lumsden and Hall 1975a)

algae (eucaryotes). This interesting transition is now being investigated to explore the evolution from algae to higher plant chloroplasts, and obviously it is very interesting to study the evolutionary lines from the green algae to higher plant-type chloroplasts. The higher plant chloroplasts do, however, contain a cyanide-insensitive, membrane-bound SOD which may also occur in blue-green algae (Lumsden and Hall, 1975a). This is rather a complicated system, but we think that the membrane-bound SOD may be a Mn-containing enzyme and would have been simultaneously involved in the development of the water-splitting reaction which gave rise to O_2 evolution and the protection of the membrane against O_2 toxicity. This possibility is now being actively investigated.

The reason that less-developed photosynthetic organisms and also the Clostridial-type bacteria probably used Fe or Mn in the active center of SOD instead of Cu, as higher organisms do, probably results from the fact that under anaerobic conditions, Cu is highly insoluble and would not have been available to form enzymes (Egami, 1975; Osterberg, 1974). With the development of O_2 in the atmosphere, which then became more oxidizing, Cu probably became more soluble and would thus be available for complexing into various types of protein. Cu proteins are quite widespread in aerobic organisms and probably function just as efficiently in SOD as do the cyanide-insensitive Fe- and Mn-containing SODs, which, to a large extent, they seem to have replaced in higher aerobic organisms.

REFERENCES

Allen, J. F., and Hall, D. O. 1974. *Biochem. Biophys. Res. Comm.,* *58*:579.

Asada, K., Yoskihawa, K., Takahadi, M., Maeda, Y., and Enmayi, K. 1975. *J. Biol. Chem., 250*:2801.

Boucher, F., and Gingras, G. 1975. *Biochem. Biophys. Res. Comm.,* *67*:421.

Bridgen, J., Harris, J. I., and Northrop, F. 1975. *FEBS Lett.* *49*:392.

Broda, E. 1975. *Evolution of the Bioenergetic Processes.* Pergamon Press, Oxford.

Carr, N. G., and Whitton, B. A. (eds.). 1973. *Biology of Blue-green Algae.* Blackwell, Oxford.

Egami, F. 1975. *J. Biochem.* (Tokyo),*77*:1165.

Fridovich, I. 1975. *Ann. Rev. Biochem., 44*:147.

Hall, D.O., Cammack, R., Rao, K. K., Evans, M. C. W., and Mullinger, R. N. 1975a. *Biochem. Soc. Trans., 3*:361.

Hall, D. O., Rao, K. K., and Cammack, R. 1975b. *Sci. Prog. Oxf.,* *62*:285.

Hall, D. O., Rao, K. K., and Mullinger, R. N. 1975c. *Biochem. Soc. Trans., 3*:472.

Halliwell, B. 1974. *New Phytol., 73*:1075.

Hewitt, J., and Morris, J. G. 1975. *FEBS Lett., 50*:315.

Holm, R. H. 1975. *Endeavour, 34*:38.

Lavelle, F., Durosay, P., and Michelson, A. M. 1974. *Biochimie, 56*:451.

Lindmark, D. G., and Müller, M. 1974. *J. Biol. Chem 249*:4634.

Lovenberg, W. (ed.). 1973. *Iron - Sulfur Proteins, Vols. I and II. Academic Press, New York.*

Lumsden, J.,*1975. Ph. D. thesis, University of London King's College.*

Lumsden, J., and Hall, D. O. 1975a. *Biochem. Biophys. Res. Comm., 64*:595.

Lumsden, J., 1975b. *Nature (London), 257*:670.

Morris, J. G. 1975. *Adv. Microb. Physiol. 12*:169.

Muir, M. D., and Hall, D. O. 1974. *Nature (London) 252*:376.

Osterberg, R. 1974. *Nature (London) 249*:382.

Schopf, J. W. 1975. *Endeavour, 34*:51.

Tel-Or, E., Cammack, R., and Hall, D. O. 1975. *FEBS Lett. 53*:135.

Tel-Or, E., Cammack, R., and Hall, D. O. 1976. *Biochem. Biophys. Acta,* (in press).

22

LOW OXYGEN LEVELS AND THE PALMITOYL CoA DESATURASE OF YEAST: RELATION TO PRIMITIVE BIOLOGICAL EVOLUTION

NANCY SYMMES WHITAKER and HAROLD P. KLEIN
Ames Research Center, NASA

The present high levels of oxygen in the atmosphere are believed to have arisen as a result of plant photosynthesis, possibly 2 to 3 billion years ago. Life as it evolved on earth branched out into two groups: *Prokaryotes* and *Eukaryotes*. The more complex organisms, the *Eukaryotes*, are thought to have arisen in the period after high levels of oxygen became available, largely on the basis that they are almost all "aerobic" organisms. We wish here to point out that, in addition to respiratory activity, these organisms also carry out a number of other oxygen-requiring processes; and pose the question whether certain of these well known aerobic processes which commonly occur in *Eukaryotes* could have earlier origins, i.e., whether they arose at a time when concentrations of oxygen were very low. Berkner and Marshall (1965), for example, have suggested that the primitive atmosphere contained small amounts of oxygen, of the order of 0.001 of present atmospheric levels (PAL) of oxygen.

One such eukaryotic organism, the yeast, *Saccharomyces cerevisiae*, has been extensively studied as to its sensitivity to changes in oxygen concentration, since the early work of Pasteur (1861). In addition to oxidative respiration, yeasts carry out many other oxygen-requiring processes, such as the biosynthesis of sterols and of unsaturated fatty acids. Using this yeast, Rogers and Stewart (1973) have investigated the oxygen levels necessary *in vivo* for many of these processes. They found that the respiratory rate in their strain of yeasts was half-maximal at oxygen concentrations of 0.01 to 0.08 PAL, whereas the cellular content of sterols and unsaturated fatty acids was already at maximal levels in this range of oxygen concentration. They also reported that the half-maximal oxygen concentration for a full complement of cytochrome oxidase, sterols, and unsaturated fatty acids was 0.002 PAL.

Our laboratory has been studying various aspects of lipid metabolism in this yeast. In these studies we observed long ago (Klein, 1955) that when cells are grown in "standing" cultures, during which they metabolize by fermentation rather than respiration, the yeast cells contain moderate amounts of both unsaturated fatty acids and sterols, both of which require oxygen for their synthesis. Similar results were obtained whether the cells were grown in a relatively complex medium, or in a completely synthetic medium devoid of added fatty acids or sterols. We have not measured the oxygen concentration in such fermenting cultures, but David and Kirsop (1973), have reported concentrations of 3 to 12 M for such solutions, using several strains of malt yeast. This is to be compared to a concentration of 230 M at air saturation at 30°C.

Since oxygen-requiring lipid synthesis was occurring at apparently low oxygen concentrations in our yeast, we turned our attention to one of the key enzymes in this process, palmitoyl-CoA desaturase. The key enzyme system involved in this reaction was first described by Bloomfield and Bloch (1960), who reported that desaturation of palmitoyl-CoA to palmitoleyl-CoA required oxygen and NADPH. The absolute dependence of the reaction on oxygen is demonstrable only when oxygen is girorously excluded from all solutions. For the experiment summarized in Table I, the buffers were boiled and placed immediately into a nitrogen-flushed glove box containing a methylene blue anaerobic indicator. Enzyme assays were carried out inside this arraratus using large diameter test tubes as the reaction vessels. As is seen, no activity is present under these conditions over a wide range of substrate concentrations.

Having established that there was no *anaerobic* desaturase activity in these cells, and that the measured activity showed an absolute requirement for oxygen, it then became possible to set up experiments in which the activity of the desaturase could be measured as a function of oxygen concentration. For these, all manipulations were carried out inside the glove box described above. Enzyme and substrate were loaded into tubes which were connected to a vacuum stopcock and closed to the atmosphere. These tubes were then transferred to a vacuum manifold and known gas mixtures were introduced. The enzyme and substrate were combined and incubated for 8 minutes. The results from such an experiment are shown in Figure 1. Here, we have plotted specific activity of the enzyme as a function of the oxygen concentration expressed as percent in the atmosphere.

TABLE I
*Desaturase Activity in the Presence and Absence of Oxygen**

^{14}C-Palmitoyl-CoA (nmoles)	Percent Conversion to Palmitoleyl-CoA		
	Aerobic Incubation (a)	Anaerobic Incubation (a)	(b)
6	22.1	0.4	0.0
11	24.7	0.0	0.0
19	23.6	0.0	0.0
36	22.8	0.0	0.0

*Crude mitochondrial preparation, suspended in Tris-Mg buffer
(2mM, pH 7.5), was added to solution containing 1-^{14}C-Palmitoyl-
CoA, NADPH (0.5 moles), and phosphate buffer (100 moles, pH 7.0)
in a total volume of 1.0 ml. Enzyme preparation contained: a)
100 g protein; b) 200 g protein. Incubation was at 30° C. for
8 minutes after which fatty acids were extracted, quantitated by
gas chromatography, and counted in a scintillation counter. Anaerobic incubation was in a glove box; aerobic incubation was
carried out by exposing the enzyme preparation and the substrate
solution to air in glove box for 15 minutes at room temperature
prior to combining them, and incubating the complete system for
8 minutes in air. Total radioactivity per assay was 350,000 cpm.

From the data, it is clear that the activity of palmitoyl-
CoA desaturase appears to be close to saturated with respect to
oxygen at very low concentrations of oxygen. The "apparent"
half-maximal activity is reached at an oxygen concentration of
about 0.04 percent of the atmosphere. This is an oxygen concentration 500-fold lower than that of present atmospheric level of
oxygen, and is in the range of oxygen concentration which Rutten
(1970) suggests was present in the atmosphere 3 to 4 billion years
ago.
     Bloch (1968) has suggested that the desaturation of fatty
acids which occurs by an oxygen requiring process is less primitive than the anaerobic pathway, and organisms which contain the
aerobic pathway are more recently evolved. Our preliminary data
offered here suggest that enzymatic processes which require oxygen, such as those involved in the formation of yeast lipids,
could have occurred under conditions postulated to have existed
on the primitive earth prior to the emergence of an oxygen-rich
atmosphere. It appears that the oxygen requirement for the desaturase has no bearing as to the time of evolution of yeast and

⁺hat the desaturase pathway could have more primitive origins than is currently assumed.

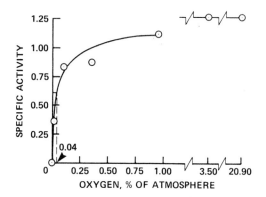

Fig. 1. Activity of Palmitoyl-CoA Desaturase at Different Oxygen Concentrations. Tubes contained 6 nmoles substrate and 100 g protein; otherwise conditions as in Table I.

REFERENCES

Berkner, L. V., and Marshall, L. C., 1965, *J. Atmos. Sci.* 22, 225.
Bloch, K., 1968, *Acc. Chem. Res.* 2, 193.
Bloomfield, D. K., and Bloch, K., 1960, *J. Biol. Chem.* 235, 337.
David, M. H., and Kirsop, B. H., 1973, *J. Gen. Microbiaol.* 77, 529-531.
Klein, H. P., 1955, *J. Bacteriol.* 69, 620.
Pasteur, L., 1861, *Compt. rend.* 52, 1260.
Rogers, P. J., and Stewart, P. R., 1973, *J. Bacteriol.* 115, 88-97.
Rutten, M. G., 1970, *Space Life Sci.* 2, 5-17.

Subject Index

A

Acanthomorphitae, 159
Accretion, 2, 3, 5–6, 14
Albedo
  andesites, 15
  ice cover, 16
Algal mats, 113–115, 133–140
Alluvial fans, 17
Amelia Dolomite, 115
Amino acids, 21, 101
Amitosoq Gneiss, 25
Ammonium ion exchange in clays, 15–16
Animikiea septata Barghoorn, 125–126
Aphonothece, 198
Arkoses sediments, 18, 35
Atmosphere
  backscattering, 15
  cloud cover, 15
  composition, 2, 16, 88–90
  degassing, 1–11, 88–90
  dissipation, 2, 3
  dust, 15
  formation, 3, 6, 87–98
  origin, 1–11, 14
  oxygen budget, 87–100
  pressure, 14
  Venus, 3–4

B

Bacillus stearothermophilis, 198, 200
Back River, 44
Baicalia burra, 114
Ballantrae Group, 155, 158, 159
Barney Creek Shale, 83
Basalts, 7, 15–16
  Fossa Magma, 51
  magma types, 33
  oceanic, 29–34

Slave Province, 41, 44
  (*see* Volcanoes)
Beck Spring Dolomite, 115
Belcher Group, 115
Belingwe Stromatolites, 102
Biotite, 36
Bitter Springs Formation, 115, 163–164
Blue-Green Algae
  algal mats, 113–115, 133–140
  photoreactivation, 137
  silicification, 181–183
  silicates, 20–22
  ultraviolet protection, 135–140
  (*see* Gunflint, *Lyngbya*, Stromalites)
Bulawayan Stromatolites, 88, 102, 134, 171
Bungle Bungle Dolomite, 115

C

Canadian Shield, 41–54
  formation, 29
  graywackes, 34
  Grenville Province, 29
  Llano uplift, 30
  reworking, 29
  Swaziland Supergroup comparison, 31
  (*see* Slave Province)
Carbonates
  biological weathering, 14
  evaporites, 19
  organic carbon, 55
  nonbiogenic, 15, 18–19
  stromatolites, 20–22, 112–113
  surf zone, 19
  Urey equilibrium, 15, 18–19
  Yellowknife sediments, 42, 46
Carbon cycle, 88–90
  isotopic balance, 90–94
  reservoir, 94–98
Carbon Dioxide